# Das überzeugende Vorstellungsgespräch auf Englisch

*Christian Püttjer* und *Uwe Schnierda* arbeiten seit 1992 als Trainer und Berater in den Bereichen Karriere, Bewerbung und Rhetorik. Ihre Erfahrungen aus Bewerbungsmappen-Checks, Einzelberatungen und Seminaren haben sie, angereichert durch viele Tipps und Übungen, in zahlreichen Ratgebern veröffentlicht. Bei Campus erscheinen von Püttjer & Schnierda unter anderem *Handbuch Einstellungstest*, *Trainingsmappe Einstellungstest Allgemeinbildung* und *Trainingsmappe Assessment-Center*.

*Dr. Steve Williams* ist britischer Staatsbürger und Germanist, arbeitet als Übersetzer und hat bereits mehrere Englischarbeitsbücher auf Deutsch für verschiedene Zielgruppen verfasst.

Christian Püttjer & Uwe Schnierda
unter Mitarbeit von Dr. Steve Williams

# Das überzeugende Vorstellungsgespräch auf Englisch

Die 200 entscheidenden Fragen und
die besten Antworten

Campus Verlag
Frankfurt/New York

Bibliografische Information der Deutschen Nationalbibliothek:
Die Deutsche Nationalbibliothek verzeichnet diese Publikation in der
Deutschen Nationalbibliografie. Detaillierte bibliografische Daten
sind im Internet unter http://dnb.d-nb.de abrufbar
ISBN 978-3-593-38701-7

Copyright © 2009 Campus Verlag GmbH, Frankfurt/Main
Umschlaggestaltung: grimm.design, Düsseldorf
Satz: Publikations Atelier, Dreieich
Druck und Bindung: AALEXX Buchproduktion GmbH, Großburgwedel
Gedruckt auf säurefreiem und chlorfrei gebleichtem Papier.
Printed in Germany

Besuchen Sie uns im Internet: www.campus.de

# Inhalt

# Englisch: die neue Herausforderung im Job-Interview

Immer häufiger erreichen uns in unserer Beratungspraxis Anfragen von Kunden, die sich auf Job-Interviews in englischer Sprache vorbereiten wollen. In Zeiten globalisierter Arbeitsprozesse ist dies auch kaum verwunderlich. Einige unserer Kunden wollen sich bei deutschen Tochterunternehmen US-amerikanischer Konzerne bewerben. Andere möchten für asiatische Konzerne in Europa tätig werden. Wiederum andere streben eine Position im Ausland an. Und dann gibt es auch noch Unternehmen in Deutschland, die sich für Englisch als Geschäftssprache entschieden haben und deshalb bei ihrer Bewerberauswahl englische Job-Interviews einsetzen.

## Warum werden englische Job-Interviews in Deutschland eingesetzt?

Job-Interviews auf Englisch haben in den letzten Jahren stark zugenommen. Betraf dies früher hierzulande überwiegend (deutschsprachige) Bewerber, die in den USA, in Großbritannien, in Kanada, Australien oder Neuseeland arbeiten wollten, ist es mittlerweile anders geworden. Die ursprüngliche Gruppe der Auslandsbewerber gibt es natürlich immer noch. Aber zusätzlich gibt es heutzutage eine weitere Gruppe von Bewerbern, die sich englischen Job-Interviews stellen muss, allerdings direkt in Deutschland oder Europa. Festzuhalten bleibt also, dass der Einsatz der englischen Sprache bei der

Personalauswahl in dem Maße zugenommen hat, in dem die Personalgewinnung internationaler geworden ist.

Europaweit tätige Personalberatungen führen daher Auswahlgespräche mit deutschen Kandidaten auf Englisch. Auch international tätige deutsche Unternehmen wollen sicherstellen, dass zukünftige Mitarbeiter sich auf Englisch verständigen können. Tochterunternehmen amerikanischer Konzerne, die in Deutschland angesiedelt sind, benutzen zwar im Arbeitsalltag häufig die deutsche Sprache, bei direkten Kontakten zum US-Headquarter oder bei internationalen Meetings ist dann aber ebenfalls Englisch gefragt. Da also Englisch im Arbeitsalltag eine immer größere Rolle spielt, werden mittlerweile englische Job-Interviews in Deutschland viel häufiger als früher eingesetzt.

### Ein doppelter Stresstest wartet auf Sie

Englische Job-Interviews sind eine echte Doppelbelastung für die Bewerberinnen und Bewerber. Schließlich sind Vorstellungsgespräche ohnehin schon stressvoll genug. Es gilt, die eigenen Stärken darzustellen, Anforderungen des Unternehmens zu erfüllen, Beispiele für erfolgreiches Arbeiten zu geben und persönliche Soft Skills wie Teamfähigkeit, Kommunikationsstärke und Belastbarkeit aussagekräftig zu vermitteln. Um in Vorstellungsgesprächen punkten zu können, sollte man auch die Gesprächsstrategie der Firmenseite durchschauen, im richtigen Augenblick die passenden Argumente bringen und sich nicht verunsichern lassen. Das alles ist schon einmal leichter gesagt als getan, aber nun kommt als zusätzlicher Belastungsfaktor noch hinzu, dass Sie als Bewerber in eine fremde Sprache umschalten müssen. Wie erläutern Sie auf Englisch Ihre berufliche Entwicklung? Wie präsentieren Sie Ihre Stärken in dieser Fremdsprache? Und wie machen Sie in englischer Sprache deutlich, dass Sie sich gründlich über Ihr künftiges Arbeitsfeld und die dazugehörigen Aufgaben informiert haben? Während Sie sich Ihre Antworten gedanklich – auf Deutsch – zurechtlegen, müssen Sie zeitgleich überlegen, wie Sie das Ganze auch noch in englischer Sprache ausdrücken. Und dieser doppelte Stresstest will erst einmal gemeistert werden.

## Überzeugen Sie auch auf Englisch

An dieser Stelle setzt unser Praxisratgeber an. Wir haben für Sie Hunderte englische Beispielformulierungen zusammengestellt, die sich an den Erfordernissen von Job-Interviews orientieren. Weder handelt es sich dabei um eine trockene Grammatik noch um eine reine Vokabelsammlung. Ganz im Gegenteil lernen Sie in praxisnahen Zusammenhängen, welche Antworten Sie auf die am häufigsten in Vorstellungsgesprächen gestellten Fragen geben können. Und zwar in englischer Sprache. Denn genauso, wie es ein technisches Englisch oder ein Business-Englisch gibt, gibt es auch ein »Karriere«-Englisch. Und dieses »Karriere«-Englisch ist immer dann unverzichtbar, wenn es um Ihr berufliches Vorwärtskommen geht.

Da Sie bei englischen Job-Interviews unter der geschilderten doppelten Stressbelastung stehen, geben wir Ihnen auch eine doppelte Hilfestellung. Sie lernen zum einen, sich auf die Wünsche von Personalverantwortlichen, künftigen Fachvorgesetzten und anderen Entscheidungsträgern auf der Firmenseite einzustellen. So ist es beispielsweise gerade in englischen Job-Interviews wichtig, eine Hands-on-Mentalität zu vermitteln und Ihren persönlichen Anteil an Firmenerfolgen deutlich herauszustellen. Zum anderen geben wir Ihnen Formulierungen an die Hand, mit denen Sie sich in englischer Sprache positiv in Szene setzen können. Dabei haben wir zusammen mit unserem Sprachexperten Dr. Steve Williams auf ein allgemein verständliches und nachvollziehbares Englisch Wert gelegt. Schließlich treten Sie nicht als Muttersprachler, sondern als deutschsprachiger Fachspezialist beziehungsweise als deutschsprachige Führungskraft auf. Präsentieren Sie sich als Macher in Ihrem Fachgebiet, der sein Können auch in englischer Sprache überzeugend vermitteln kann. Wie dies im Einzelnen geht, veranschaulichen wir Ihnen mit unserem Trainingsprogramm.

# Bewerben mit der Püttjer & Schnierda-Profil-Methode

Gesichtslose Bewerber, die austauschbar erscheinen, machen es sich und den Firmen unnötig schwer, zueinanderzufinden. Machen Sie es besser: Sie werden sich im Bewerbungsverfahren mehr Aufmerksamkeit verschaffen, wenn Sie Ihr Profil aussagekräftig und glaubwürdig vermitteln können.

Die Profil-Methode, die wir dazu in unserer über 15-jährigen Beratungspraxis entwickelt haben, hat schon vielen Bewerbern zu mehr Erfolg verholfen (www.karriereakademie.de).

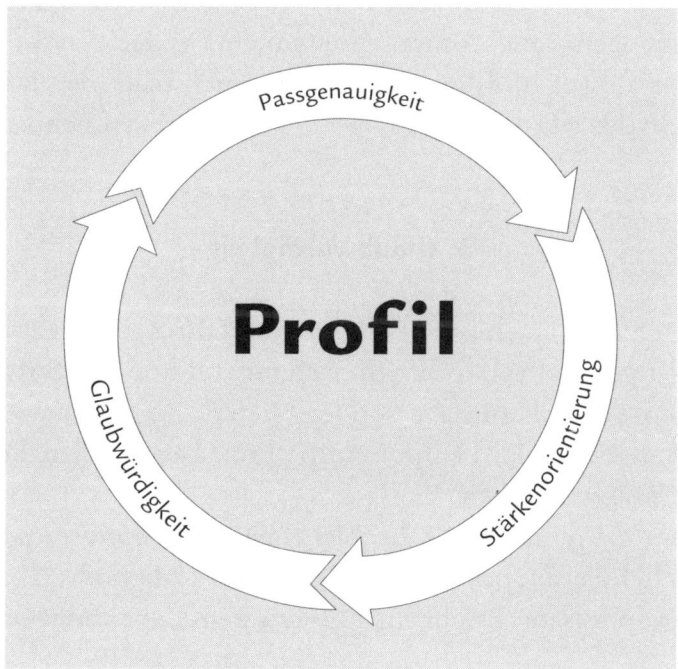

Drei Kernelemente kennzeichnen die Profil-Methode: Punkten Sie mit einer passgenauen Bewerbung, vermitteln Sie Ihre Stärken, und treten Sie glaubwürdig auf.

## 1. Passgenauigkeit

Je besser Sie im Bewerbungsverfahren auf die Anforderungen des Berufes eingehen, desto höher ist Ihre Erfolgsquote. Machen Sie sich den Blick der Personalverantwortlichen zu eigen. Argumentieren Sie von den Anforderungen der zu vergebenden Stelle her. So wird Ihr Auftritt passgenau.

## 2. Stärkenorientierung

Niemand lässt sich durch Krisen- und Problemschilderungen von etwas überzeugen – auch Unternehmen nicht! Verzichten Sie deshalb auf Abwertungen und Relativierungen, und stellen Sie lieber Ihre Vorzüge in den Mittelpunkt. So werden Ihre Stärken sichtbar.

## 3. Glaubwürdigkeit

Verbiegen Sie sich nicht im Bewerbungsverfahren, denn Ihre Persönlichkeit ist gefragt! Verstecken Sie sich nicht hinter Leerfloskeln und abstrakten Formulierungen, sondern liefern Sie stattdessen nachvollziehbare Beispiele, die Ihren Auftritt mit Leben füllen. So gewinnen Sie Glaubwürdigkeit.

Alle im Campus Verlag erschienenen Bewerbungsratgeber von Püttjer & Schnierda basieren auf der Profil-Methode. Profitieren auch Sie von unserer Erfahrung und unserem Expertenwissen!

# 1

# Welche Gesprächssituationen erwarten Sie?

Bevor es an nun an das eigentliche Trainingsprogramm geht, möchten wir Sie noch für die Gesprächssituationen, in denen Sie Ihr »Karriere«-Englisch einsetzen können, sensibilisieren. Sie werden sehen, dass Sie Ihre neu erworbenen und ausgebauten Sprachkenntnisse nicht nur in klassischen Vorstellungsgesprächen werden nutzen können. Es gibt noch viele weitere Anlässe, bei denen es sich gut macht, einige Sätze zum eigenen beruflichen Hintergrund auf Englisch sagen zu können.

## Vom klassischen Job-Interview bis zum Social-Networking

Die Kunden, die wir in unserer Beratungspraxis beim Erreichen ihrer Karriereziele oder direkt zur Vorbereitung auf Job-Interviews coachen, berichten von diesen Gesprächssituationen, in denen ihr Englisch gefragt war:

- klassisches Job-Interview auf Englisch (Vorstellungsgespräch),
- klassisches Job-Interview auf Deutsch (Vorstellungsgespräch),
- telefonisches Job-Interview,
- Kontaktgespräch mit einem Headhunter,
- Interview im Assessment-Center,
- Selbstpräsentation im Assessment-Center,
- Career-Networking (Tagungen, Messen, Geschäftsessen, Geschäftskontakte).

**Klassisches Job-Interview (Vorstellungsgespräch):** Diese Gesprächssituation, das klassische Vorstellungsgespräch auf Englisch, verlangt von Ihnen die größten Anstrengungen in Sachen »Karriere«-Englisch. Schließlich erwarten Sie zahlreiche Fragen zu Ihrem beruflichen Hintergrund: Über welche Kenntnisse und Fähigkeiten verfügen Sie? Wo liegen Ihre Stärken? In welchen Arbeitsbereichen kennen Sie sich aus? Und was haben Sie bisher beruflich erreicht? Wer Führungsaufgaben anstrebt, wird darüber hinaus auch mit Fragen zum Führungsverständnis und zur Mitarbeitermotivation konfrontiert. Ein komplettes Vorstellungsgespräch kann – je nach zu besetzender Position – durchaus zwei Stunden und länger dauern. Oftmals werden auch mehrere Job-Interviews mit Ihnen geführt, bis es zu einer Einstellungszusage kommt.

**Klassisches Job-Interview auf Deutsch:** Nicht immer können Sie sich in Sicherheit wiegen, wenn das Vorstellungsgespräch auf Deutsch beginnt. Unsere Kunden berichten auch von Vorstellungsgesprächen, in die ein Kurztest der englischen Sprachkenntnisse gleich integriert wird. Dann findet zwar der Großteil des Interviews auf Deutsch statt, doch plötzlich bittet man den Bewerber, doch einmal seinen Werdegang in englischer Sprache zu schildern, denn wie man seinen schriftlichen Unterlagen entnehmen könne, verfüge er ja über sehr gute Englischkenntnisse. Auch einige Nachfragen zur Selbstdarstellung in Englisch sind dann möglich und müssen ebenfalls in Englisch beantwortet werden.

**Telefonisches Job-Interview:** Das telefonische Job-Interview wird von Personalberatungen und Unternehmen gerne genutzt, um sich einen ersten Eindruck über Bewerber zu verschaffen. Üblicherweise werden telefonische Job-Interviews vorab angekündigt, und es erfolgt meistens ein Hinweis darauf, dass das Gespräch auf Englisch geführt wird. Diese Interviews sind kürzer als persönliche Job-Interviews, dennoch gilt es auch hier, zu überzeugen. Gerade aufgrund der knappen Zeit und der fehlenden visuellen Rückmeldung müssen Sie mit Ihren Antworten im telefonischen Job-Interview schnell auf den Punkt kommen. Suchen Sie erst im Telefongespräch nach den richtigen englischen Ausdrücken und Formulierungen, wird das Gespräch schneller, als Ihnen lieb ist, enden.

**Kontaktgespräch mit einem Headhunter:** Das Geschäft der Personalberatungen ist immer internationaler geworden, daher verwundert es nicht, dass der Executive Search durch sogenannte »Headhunter« auch immer häufiger auf Englisch stattfindet. Bei der ersten Kontaktaufnahme durch international tätige Personalberatungen handelt es sich natürlich nicht gleich um ein in die Tiefe gehendes Job-Interview. Angesprochene Kandidatinnen und Kandidaten sollten jedoch einige wichtige Stichworte zu ihrem Profil auf Englisch liefern können. Im weiteren Verlauf sind ebenfalls Ihre Englischkenntnisse gefragt, um Ihre beruflichen Vorstellungen mit den Angeboten der Headhunter abzugleichen. Ein Muss ist Englisch ebenfalls, wenn Sie eine – deutsche – Personalberatung ansprechen, um internationale Aufgaben zu übernehmen. Zeigen Sie schon in den Kontaktgesprächen mit Headhuntern, dass Sie über sichere Englischkenntnisse verfügen.

**Interview im Assessment-Center:** Assessment-Center sind Personalauswahlverfahren, die aus mehreren Übungseinheiten wie beispielsweise Gruppendiskussionen, Präsentationen und Rollenspielen bestehen und vorrangig Ihre Soft Skills wie Teamfähigkeit, Belastbarkeit, Entscheidungsfreude oder Führungskompetenz überprüfen sollen. Wenn Sie sich vertiefend mit dem Thema Assessment-Center beschäftigen möchten, empfehlen wir Ihnen unsere *Trainingsmappe Assessment-Center. Die häufigsten Aufgaben – die besten Lösungen.* Auch in Assessment-Centern können gute Kenntnisse in »Karriere«-Englisch eine wichtige Rolle spielen. Denn in manchen Assessment-

Centern werden auch Interviews in englischer Sprache durchgeführt.

**Selbstpräsentation im Assessment-Center:** Zu Beginn der meisten Assessment-Center wartet eine echte Herausforderung auf die Teilnehmerinnen und Teilnehmer: die Selbstpräsentation. In Selbstpräsentationen geht es um eine kurze Darstellung des beruflichen Werdeganges und der eigenen Stärken. Wenn die Selbstpräsentation in englischer Sprache gewünscht wird, müssen Sie hellwach sein, schließlich geht es um den wichtigen ersten Eindruck bei den AC-Beobachtern, die am Ende des Auswahlverfahrens über eine Einstellung entscheiden. Präsentieren Sie sich gleich zu Anfang mit holprigem Englisch, haben Sie bereits den Start verpatzt. Wer es besser machen will, sorgt vor und entwirft bereits zu Hause seine englische Selbstpräsentation. Und die sollte genauso aufgebaut sein wie die Antwort auf die typische Einstiegsfrage in Job-Interviews: *Why should we give you the job?*

**Career-Networking (Tagungen, Messen, Geschäftsessen, Geschäftskontakte):** Es gibt im beruflichen Kontext noch viele weitere Möglichkeiten, sich selbst und seine beruflichen Erfolge ins Gespräch zu bringen. Viele Jobangebote kommen über Kontakte und Empfehlungen zustande. Wer sich am Rand von Workshops, Seminaren oder Tagungen aktiv um die Pflege und den Aufbau beruflicher Kontakte kümmert, hat es leichter, seinen Wunschjob zu bekommen. Finden diese Veranstaltungen in einem internationalen Rahmen statt, ist es natürlich unabdingbar, auf Englisch etwas zum eigenen beruflichen Hintergrund sagen zu können. Darüber hinaus sind auch Geschäftsessen und Kontakte zu geschäftlichen Partnern wie Kunden und Lieferanten interessant für die Karriereplanung. Und auch hier gilt: Je internationaler die Geschäftspartner sind, desto unverzichtbarer ist Ihr »Karriere«-Englisch.

## Wer sitzt Ihnen gegenüber?

Ihre Gesprächspartner auf der Firmenseite werden in erster Linie Personalverantwortliche, zukünftige Fachvorgesetzte, Geschäftsführer oder Headhunter der Personalberatungen sein. Sie können aber auch auf Betriebsräte oder Personalratsmitglieder treffen. Beim Networking tauschen Sie sich mit Kollegen aus anderen Unternehmensbereichen und Ansprechpartnern und Geschäftspartnern aus anderen Unternehmen aus. Wichtig ist es, alle Anwesenden gleichermaßen ernst zu nehmen, und zwar sowohl bei offiziellen Personalauswahlverfahren als auch bei informellen Treffen. Es darf Ihnen nicht passieren, dass Sie sich ausschließlich auf den Wortführer konzentrieren und die restlichen Anwesenden kaum eines Blickes würdigen. Trainieren Sie also, parallel zu Ihren Antworten den Blickkontakt zu allen Gesprächspartnern zu suchen.

Wenn Sie unsere positiven Beispielantworten aufmerksam lesen und auswerten, werden Sie feststellen, dass die vorgestellten Strategien auf alle Gesprächspartner gleichermaßen zugeschnitten sind. Für Personalexperten stehen dabei eher nichtfachliche Fähigkeiten wie Teamfähigkeit, die Fähigkeit zur Selbstmotivation, Konfliktfähigkeit und Kommunikationsfähigkeit im Vordergrund. Künftige Fachvorgesetzte legen hingegen Wert darauf, zu erfahren, ob Sie über die entsprechenden Fach- und Branchenkenntnisse verfügen. Und Geschäftsführer sind besonders daran interessiert, festzustellen, wie lange es dauert, bis die Firma von Ihrer Mitarbeit profitiert, und ob Sie Arbeitsabläufe optimieren oder Kosten senken können. Orientieren Sie sich deshalb an den Beispielantworten, um im Vorstellungsgespräch allen Vorlieben und Interessen gerecht zu werden.

Je nach Größe des Unternehmens und nach den Vorlieben der Entscheidungsbeteiligten müssen Sie mit einem eher strukturierten Gespräch oder einem freien Gespräch rechnen. Strukturierte Einstellungsinterviews werden vor allem in großen Konzernen eingesetzt, um das Auftreten der Kandidaten besser miteinander vergleichen zu können. In kleinen Firmen und bei Kontaktgesprächen mit Headhuntern geht es freier zu. Aber auch hier gibt es bestimmte Fragen, die unbedingt beantwortet werden müssen. Obwohl es also Unterschiede im Ablauf gibt, gilt dennoch bei allen Auswahlverfahren, dass Ihre Antworten auf Englisch nur dann überzeugen werden, wenn sie ausreichend Informationen, Argumente und Beispiele enthalten.

# 2

# Welche Fragen werden Ihnen gestellt?

Bevor Sie nun gleich in den Praxisteil einsteigen, möchten wir Ihnen noch einen Überblick darüber geben, aus welchen Themenbereichen die Fragen stammen, die Sie im Job-Interview erwarten. Schließlich ist dieses Trainingsbuch als Praxisratgeber konzipiert, und das bedeutet für Sie, dass wir die Fragen so zusammengestellt haben, wie sie unsere Kunden auch bei Auswahlgesprächen in Unternehmen erlebt haben. Außerdem können Sie so besser erkennen, mit welcher Zielrichtung Ihnen einzelne Fragen gestellt werden. Dieses Wissen hilft Ihnen dabei, Gesprächsabsichten der Firmenseite schneller zu erkennen und taktisch angemessen zu reagieren.

## Die wichtigsten Fragenkomplexe im Überblick

Die folgende Übersicht zeigt Ihnen die verschiedenen Themenbereiche, die wir mit Ihnen in dieser Trainingsmappe durchgehen. Es erwarten Sie Fragen aus diesen Bereichen:

*Fragen zur beruflichen Qualifikation:*
- Why should we give you the job? (Warum sollten wir gerade Sie einstellen?)
- What can you do for us? (Was können Sie für uns leisten?)
- Are you customer-oriented? (Verfügen Sie über Kundenorientierung?)
- How good are your PC skills? (Wie gut sind Ihre PC-Kenntnisse?)

*Fragen zum Unternehmen:*
- What do you know about our company? (Was wissen Sie über unsere Firma?)

*Fragen zur persönlichen Qualifikation:*
- How do you cope with change? (Wie gehen Sie mit Veränderungen um?)
- How do you motivate yourself for work duties? (Wie motivieren Sie sich für berufliche Aufgaben?)
- Do you have a realistic self-image? (Ist Ihr Selbstbild realistisch?)
- How do you deal with conflict? (Kennen Sie Ihr Konfliktverhalten?)

*Stressfragen und Vorurteile:*
- How do you deal with stress questions and unlawful questions? (Wie entschärfen Sie Stressfragen und unzulässige Fragen?)
- Are you able to dispel prejudice? (Können Sie Vorurteile entkräften?)

*Fragen zur Führungserfahrung:*
- What kind of people manager are you? (Wie führen Sie Ihre Mitarbeiter?)

*Fragen zur Gehaltsvorstellung:*
- What are your salary expectations? (Welche Gehaltsvorstellungen haben Sie?)

*Fragen im zweiten Vorstellungsgespräch:*
- What can you expect in the second interview? (Was erwartet Sie im zweiten Vorstellungsgespräch?)

*Eigene Fragen:*
- What questions should you ask? (Welche Fragen sollten Sie stellen?)

Insgesamt haben wir für Sie 200 Beispielfragen aus Job-Interviews zusammengestellt. Die dazugehörigen 400 Beispielantworten – jeweils 200 ungeeignete und 200 geeignete – helfen Ihnen dabei, einen eigenen Stil zu entwickeln. Orientieren Sie sich an den gelungenen Antworten, aber arbeiten Sie konsequent daran, Ihr individuelles Profil in den Mittelpunkt zu stellen.

Sprachlich haben wir, da es sich ja um eine Gesprächssituation handelt, die vertrauteren Formulierungen wie zum Beispiel *didn't*, *haven't* oder *isn't* anstelle von *did not, have not* oder *is not* gewählt. Natürlich können Sie aber ebenso gut auch den etwas förmlicheren Ausdruck benutzen.

Sie können jetzt die Kapitel der Reihe nach durcharbeiten oder auch von Kapitel zu Kapitel springen. Die Struktur ist immer die gleiche: Auf der einen Seite finden Sie Fragen und Platz für Ihre eigenen Antworten, auf der folgenden Seite dann jeweils gelungene und ungünstige Beispielantworten.

# 3

# Why should we give you the job?

Fragen aus dem Themenblock *Why should we give you the job?* stehen im Mittelpunkt jedes Vorstellungsgespräches. Aus Sicht der Firma haben Bewerber hier eine Bringschuld: Sie müssen selbst begründen können, warum sie glauben, mit den Anforderungen der neuen Stelle zurechtzukommen.

## Hintergrund

Um ein Vorstellungsgespräch überhaupt in Gang zu bringen, wird der Bewerber in der Regel aufgefordert, sein berufliches Können und seinen Werdegang mit eigenen Worten zu erläutern. Die Firmenseite erwartet vor allem Informationen über die momentanen Aufgaben des Bewerbers und über besondere berufliche Erfolge. Im weiteren Verlauf des Vorstellungsgespräches wird dann mit Anschlussfragen überprüft, wie schlüssig die vorherigen Angaben des Bewerbers waren.

## Typische Fehler

Unvorbereitete Bewerber kommen nicht auf den Punkt und verlieren sich in Detailinformationen über weit zurückliegende berufliche Stationen oder die Ausbildung beziehungsweise das Studium. Oftmals wird auch eine reine Nacherzählung des Lebenslaufes geliefert –

dabei fallen zentrale Aufgaben aus der momentanen Stelle dann unter den Tisch. Es passiert auch, dass Allgemeinplätze mitgeteilt werden, ohne dass ein individuelles Profil des Kandidaten deutlich wird. Viele Bewerber begehen auch den Fehler, in ihrer Antwort keinerlei Bezug auf die Anforderungen der neuen Stelle zu nehmen.

## Negativbeispiel

Ein unvorbereiteter Bewerber wird der Aufforderung *Tell us about your career so far* häufig so nachkommen: *When I finished school, I didn't really know what I wanted to do. Luckily, my parents talked me into starting an apprenticeship. The things we had to do in training weren't always very interesting, though. I remember how we had to spend hours filing bits of metal, until one of the older apprentices came up with the idea of doing them on the lathe. Well, some of the things you have to do as an apprentice are a bit pointless ... All the same, I passed my final exam. The company couldn't give me a permanent position, but they kept me on for a while. Then I started looking for another job.*

## Kommentar zum Negativbeispiel

Hier hat ein Bewerber übersehen, dass der Personalverantwortliche vorrangig an seinem beruflichen Profil interessiert ist. Er möchte aus der Antwort heraushören können, was der Bewerber kann und ob er mit den neuen Aufgaben zurechtkommen wird. Natürlich spielt auch der Werdegang eine Rolle, allerdings nicht in dieser Breite. Es gilt, unwesentliche von wesentlichen Informationen zu trennen. Der Bewerber hätte zudem auf konkrete Tätigkeiten innerhalb der einzelnen Beschäftigungsverhältnisse eingehen müssen. So allerdings liefert er nur Anekdoten und Allgemeinplätze mit wenig Aussagekraft.

## Antwort-Strategie

Liefern Sie eine kurze Selbstpräsentation Ihres beruflichen Werdegangs, die Sie bereits zu Hause ausarbeiten und verinnerlichen sollten. Wenn Sie bereits längere Zeit im Berufsleben sind, sollten Sie sich dabei nicht in Details aus der weit zurückliegenden Ausbildung oder dem Studium verlieren. Konzentrieren Sie sich stattdessen darauf, möglichst viele Schnittpunkte zwischen Ihrer momentanen Position und der neuen Stelle herauszuarbeiten. Werden Sie konkret, indem Sie die Erfahrungen, Branchenkenntnisse und Erfolge betonen, die für die neue Stelle wichtig sind. Schließlich zeichnet sich der ideale Mitarbeiter dadurch aus, dass er ohne größere Reibungsverluste im neuen Job voll durchstarten kann.

## Positivbeispiel

Eine bessere Antwort auf die Aufforderung *Tell us about your career so far* könnte folgendermaßen lauten: *After I finished school, I decided to do an apprenticeship as an industrial mechanic. Even as an apprentice I was able to accompany installation teams on site. I also learned PLC programming with the company. When I finished my apprenticeship, I stayed with the same company for a while, then I moved over to the machine tools sector. At the moment I'm responsible for plant commissioning. Close liaison with the client and developing tailored solutions are important aspects of my work. Your job advertisement also mentioned preparing documentation and training colleagues – those are already part of my duties.*

## Kommentar zum Positivbeispiel

Der Bewerber stellt in dieser gelungenen Selbstpräsentation sehr gut Überschneidungen bisheriger Tätigkeiten mit den Aufgaben in der neuen Stelle heraus. Er weist auf seine konkreten Erfahrungen in der Inbetriebnahme hin. Es fallen die wichtigen Schlagworte Dokumentation und Schulung von Mitarbeitern, und auch seine Branchenerfahrung macht der Bewerber deutlich. Eine gute Antwort, die dem Personalverantwortlichen verdeutlicht, dass der Bewerber weiß, was auf ihn zukommt, und dass er die richtigen Kenntnisse und Erfahrungen mitbringt.

1. What made you apply for this job in particular?

Your answer: _____

_____

_____

_____

_____

_____

_____

_____

_____

_____

2. Could you summarize your background in a few sentences?

Your answer: _____

_____

_____

_____

_____

_____

_____

_____

_____

_____

_____

**Poor answer to question 1**   I read your job advertisement, and I'm very interested in the position.

**Good answer to question 1**   When I read your job advertisement, I realized it was describing me. My present duties include calculating costs and soliciting quotations. I worked on a project where we achieved better supply chain integration through the selection of suppliers. I have several years' experience in the areas of billing control, scheduling and data administration. I was particularly interested in the close liaison with field staff that you mentioned in the advertisement.

**Poor answer to question 2**   Well, after finishing Hauptschule I was unhappy with the situation, so I went back to school and did my Realschule leaving certificate. Then I did an apprenticeship as an electrical engineer. When I finished my apprenticeship, the firm didn't keep me on. I was able to get a service job with another firm. Now I'm responsible for service tasks and also have to travel a bit.

**Good answer to question 2**   After completing Realschule I decided to do an apprenticeship as an electrical engineer. Even as a trainee I took on service contracts independently. I realized that I was good at fault spotting and problem analysis in clients' systems. With my current employer I'm in charge of PLC programming for machines and preparing documentation and manuals. Also, my work includes commissioning machines for clients. I have a talent for building a good relationship with clients' operating crews, so lately I've taken over responsibility for briefing clients on site, too.

3. Please outline your job experience for us.

Your answer: _____

_____

_____

_____

_____

_____

_____

_____

_____

_____

4. Why are you sitting here today?

Your answer: _____

_____

_____

_____

_____

_____

_____

_____

_____

_____

_____

**Poor answer to question 3**   After school I wasn't sure what I wanted to do. So I went abroad for a year as an au pair, first of all. Then I worked as a sales assistant and gradually took on more and more responsibilities. Now I'm deputy manager of the branch.

**Good answer to question 3**   During my stay as an au pair in the USA I was very impressed with the American approach to retail sales. So when I got back to Germany, I did an apprenticeship as a retail saleswoman. I've always been very focused on customer service, so, for example, I reorganized the warehouse. After that, my company promoted me to deputy branch manager. Now I'm in charge of selecting product lines, inducting new employees and also sales promotions.

**Poor answer to question 4**   Well, you invited me for an interview, and I didn't want to miss out on the opportunity.

**Good answer to question 4**   I'm here, because I have experience as a supply chain manager. Alongside the usual duties, such as centralized procurement and coordinating transportation, I have been able to reduce logistic costs through just-in-time supply. I'm keen to put this experience and knowledge into practice with your company.

5. What makes you stand out from the other applicants?

Your answer: _____

_____

_____

_____

_____

_____

_____

_____

_____

_____

6. Is there a common thread to your CV?

Your answer: _____

_____

_____

_____

_____

_____

_____

_____

_____

_____

**Poor answer to question 5**   Difficult to say. Of course, I'm very aware that I'm not the only one applying. It's quite difficult to find a new job at the moment. But if you decide to pick me, you definitely won't be disappointed.

**Good answer to question 5**   I can only speak for myself, but I'd like to emphasize that I have a great deal of experience at the interface between office and field staff and between sales and marketing. I'm experienced in interpreting market research, developing marketing concepts, supporting the field sales team and also, of course, calculating offers and advising customers. I also have experience of presenting key sales data to senior management.

**Poor answer to question 6**   Unfortunately not. Somehow, things always turned out differently than I expected. I started off in personnel, then I went into PR. After a spell as a freelance lecturer I'm now looking for permanent employment again. I think it speaks for me, though, that I've got wide experience and I've never given up.

**Good answer to question 6**   The common thread is personnel support. As a personnel assistant I was mainly responsible for assigning tasks to colleagues and keeping personnel records. In PR I was in charge of internal communications, I wrote an in-house newsletter and helped to make the aims of the company clear to the workforce. After that I took on training responsibilities and contributed to human resource development plans. I'd like to bring together all of this experience now in this role as personnel officer.

7. What if this turns out to be the wrong job for you?

Your answer: _____

_____

_____

_____

_____

_____

_____

_____

_____

_____

8. Why are you interested in the advertised position?

Your answer: _____

_____

_____

_____

_____

_____

_____

_____

_____

_____

**Poor answer to question 7**   There are always risks in life. You can't say beforehand whether the new job will work out or not.

**Good answer to question 7**   I've informed myself very thoroughly about your company, and I already have several years' experience in this profession. So I'm confident that I'll adapt well to my future duties. I've always been able to establish a good relationship with my colleagues and superiors. So I'm sure that I'll be just as successful in my new position.

**Poor answer to question 8**   I'm looking forward to exciting and interesting new tasks. It's important to try something different, otherwise you get in a rut. And I'm sure that the atmosphere in the department will be better than in my current firm.

**Good answer to question 8**   Because the position is a good match with my professional experience. I enjoy working in the field. As a pharmaceutical sales rep I've always generated above-average sales in my territory. I think your new product line has great market potential. I'm keen to convince doctors, pharmacists and opinion leaders of its innovative benefits.

9. What attracts you to the new position?

Your answer: _____

_____

_____

_____

_____

_____

_____

_____

_____

_____

10. What do you see as the core of your profile?

Your answer: _____

_____

_____

_____

_____

_____

_____

_____

_____

_____

**Poor answer to question 9**   I would have been quite happy to stay with my old firm, but they went out of business. Now I'd like to carry on with your firm.

**Good answer to question 9**   I want to ensure data security and enable effective work processes. Alongside my system administration work, I've always provided a lot of user support. After all, the easier the data processing system is to use, the easier it is for employees to concentrate on their own key tasks. What really appeals to me is the chance to set up a management information system which will make the company's processes more transparent than ever before.

**Poor answer to question 10**   I'm very hard working, outgoing and open-minded. Those are the main things, but I also have plenty of professional know-how.

**Good answer to question 10**   My core tasks are financial accounting, payroll accounting, preparing company accounts, dealing with tax returns and checking tax assessments. I operate well under the pressure of a heavy workload, and I get on well with clients. I'm continually updating my knowledge of taxation law, so I'm always able to give my clients sound advice.

11. What qualifications do you have to offer us?

Your answer: _____

_____

_____

_____

_____

_____

_____

_____

_____

_____

12. How did you develop yourself professionally in your last position?

Your answer: _____

_____

_____

_____

_____

_____

_____

_____

_____

_____

_____

**Poor answer to question 11** My business administration degree, for one thing, then, of course, my managerial experience and, not forgetting, a wide range of practical experience.

**Good answer to question 11** I'm extensively qualified in the areas of product management and product line supervision. I draw up country-specific product and marketing strategies for international markets. Product positioning and pricing strategy are part of that, as is developing distribution channels. These qualifications are the fruit of many years' practical experience. At business school I learned how to carry out market and competitive analyses and to design and implement marketing strategies. Since then I've successfully put that knowledge into practice.

**Poor answer to question 12** We did courses on various software packages and went on a team-building seminar.

**Good answer to question 12** Above and beyond my secretarial duties I worked as a project assistant, so I was constantly developing my professional skills. I became thoroughly proficient in Excel, so that I could prepare numerical data. For the project presentations I learned to use PowerPoint. In a team-building seminar, that I took part in, I learned how to liaise more effectively with colleagues. I've also been studying time management outside work hours.

# 4

# What can you do for us?

Mit Fragen aus diesem Themenkomplex überprüfen und hinterfragen die Personalverantwortlichen die Arbeitserfolge der Bewerberinnen und Bewerber. Nicht das »Wollen« wie bei der Arbeitsmotivation, sondern das »Können« steht bei den Fragen zur Überprüfung der Leistungsfähigkeit im Vordergrund.

## Hintergrund

Arbeitswillige Kandidaten gibt es viele. Aus Sicht der Firmen ist es aber mindestens genauso wichtig, was bei den Anstrengungen und Bemühungen unter dem Strich herauskommt: Das Unternehmen möchte natürlich, dass der Kandidat möglichst bald gewinnbringend arbeitet. Da der Wettbewerb hart ist und die Kosten nicht aus dem Ruder laufen dürfen, sind vor allem erfolgsgewohnte Bewerber mit Kostenbewusstsein gefragt.

## Typische Fehler

Viele Bewerber haben Schwierigkeiten damit, sich mit beruflichen Erfolgen zu schmücken. Sie sind es nicht gewohnt, aus dem Dunstkreis ihres Teams herauszutreten und Beispiele dafür zu liefern, was sie persönlich dafür getan haben, um Kosten zu senken, Umsätze zu steigern, Verbesserungen einzuführen oder Qualitätsmängel zu beseitigen. Wenn aber im Vorstellungsgespräch nicht deutlich wird, welchen Anteil der Bewerber an bisherigen Abteilungs- oder Unternehmenserfolgen hatte, wird er als passiver Mitläufer abgestempelt und aussortiert.

## Negativbeispiel

Um die Leistungsfähigkeit zu überprüfen, könnte seitens der Personalverantwortlichen diese Frage eingesetzt werden: *What have you achieved so far in your areas of responsibility?* Dann darf die Antwort allerdings nicht so lauten: *I think I've always been able to keep on top of my work. There haven't really been many problems. When they have occurred, we've always solved them as a team. Our department has always achieved good results, and I've been able to make my own modest contribution to that success.*

## Kommentar zum Negativbeispiel

Hier spricht ein Bewerber, der sich unter Wert verkauft. Es ist zwar ehrbar, dass der Kandidat nicht übermäßig auftrumpfen will – leider liefert er aber nicht ein einziges Beispiel für seine erfolgreiche Arbeit. Im Gegenteil, er versteckt sich hinter der Abteilungsleistung. Eine ungeschickte Taktik, denn Personalverantwortliche könnten daraus folgern, dass der Bewerber häufiger Probleme verursacht, die sein Team dann für ihn lösen muss. Wahrscheinlich werden Personalverantwortliche auch den letzten Satz wörtlich nehmen: Sie werden vermuten, dass der Bewerber tatsächlich nur *a modest contribution*, also einen bescheidenen Beitrag, zum Abteilungserfolg beisteuern kann.

## Antwort-Strategie

Sie sollten zwar nicht als Supermann beziehungsweise Superfrau auftreten, ohne die jede Firma über kurz oder lang untergehen würde – Sie sollten sich aber daran gewöhnen, ausgewählte beruflichen Erfolge stichwortartig zu beschreiben. Überlegen Sie sich – immer mit Blick auf die Anforderungen der neuen Stelle – Beispiele aus Ihrem Tagesgeschäft. Besonders gut geeignet sind auch Erfolge aus abteilungsübergreifender Projektarbeit. Wenn es Ihnen schwerfällt, Ihre beruflichen Erfolge konkret zu benennen, können Sie auch »abgeschwächte« Formulierungen wie *I played a part in ..., I was jointly responsible for ..., Together with my colleagues I was able to ...* verwenden.

## Positivbeispiel

Um das eigene Können zu verdeutlichen, sollte die gerade schon genannte Frage *What have you achieved so far in your areas of responsibility?* besser so beantwortet werden: *I successfully introduced new products to national and international markets. To achieve this, I worked closely with the product development, production and service departments to determine country-specific marketing strategies. I managed the product positioning and price positioning, based on market and competitive analyses. I was also involved in the design and implementation of appropriate marketing strategies.*

## Kommentar zum Positivbeispiel

Diesmal tappt der Bewerber nicht in die Team-Falle. Weder relativiert er seine Leistungen, noch gibt er sich zu großspurig. Stattdessen beschreibt er nüchtern, aber aussagekräftig, welche Aufgaben ihm übertragen wurden und welche Erfolge er erzielen konnte. Es wird deutlich, dass er als Produktmanager erfolgreich tätig war und das notwendige Handwerkszeug beherrscht. Dass er sich bei seiner Arbeit eng mit Kollegen aus anderen Bereichen abstimmt, lässt er nicht unter den Tisch fallen, stellt dabei aber seinen eigenen Leistungsanteil klar heraus.

13. What are your strengths?

Your answer: _____

_____

_____

_____

_____

_____

_____

_____

_____

14. What can you do to take our company forward?

Your answer: _____

_____

_____

_____

_____

_____

_____

_____

_____

_____

**Poor answer to question 13**   I'm highly motivated, flexible and a team player.

**Good answer to question 13**   I can produce good work under pressure – for example, I was able to keep on top of day-to-day work during the changeover to a new computer system. Our customers weren't even aware of the huge restructuring task that was under way. Another of my strengths is my knowledge of different aspects of the company's work. Alongside my usual office duties I frequently took on special interdepartmental tasks like product optimisation.

**Poor answer to question 14**   I can work hard and produce good results.

**Good answer to question 14**   I'm keen to give you the benefit of my experience in interdepartmental liaison. Through discussions with colleagues I have been able to reduce processing times in my company. My keen market awareness will also be useful to you.

15. What evidence can you provide of successful work?

Your answer: _____

_____

_____

_____

_____

_____

_____

_____

_____

_____

16. What are your main tasks at the moment?

Your answer: _____

_____

_____

_____

_____

_____

_____

_____

_____

_____

**Poor answer to question 15**   It depends on what you mean. I haven't had many real failures at work, but I can't say that one thing stands out as a particular success.

**Good answer to question 15**   Firstly, I can point to my reliable performance in my daily work. Then, there's my achievement in increasing our customer base by nearly 20 per cent through direct marketing campaigns. The new product line I worked on with my project team has been a real success in the market and has given us a real competitive advantage. Even in my first job I was able to achieve improvements in customer relations through my suggestions.

**Poor answer to question 16**   At the moment I'm having to clear up after my colleagues all the time. They're always leaving things lying around, I don't really have time to get on with my own tasks.

**Good answer to question 16**   One of my main tasks is the maintenance of the production lines. I'm responsible for quick retooling when we switch products. To do that, I liaise with production planning and work closely with the service teams of the machine suppliers.

17. Which of your talents can you make use of in the advertised position?

Your answer: _____

_____

_____

_____

_____

_____

_____

_____

_____

_____

18. In your view, which are the key tasks in the new position?

Your answer: _____

_____

_____

_____

_____

_____

_____

_____

_____

_____

**Poor answer to question 17**   I don't go to pieces in a crisis, so my colleagues can always rely on me.

**Good answer to question 17**   There's my grasp of technical interrelationships, which is a big help to me in fault analysis. I get quickly to the root of the problem, and then I can make concrete recommendations for maintenance and repair. I'm good at establishing a connection with all kinds of people, so I always get on well with the clients' operating crews.

**Poor answer to question 18**   Like it says in the advertisement: you have to work well under pressure and be able to multi-task.

**Good answer to question 18**   I see the key task as customer care. As the new remit will include providing product advice, a professional manner on the phone and at trade fairs will be important. Collaboration with the field sales force is going to be another key factor. The reps will only be able to convince customers if they get good presentation material.

19. What have been your two greatest achievements?

Your answer: _____

_____

_____

_____

_____

_____

_____

_____

_____

_____


20. Have you ever made suggestions for improvement?

Your answer: _____

_____

_____

_____

_____

_____

_____

_____

_____

_____

**Poor answer to question 19**   Winning a ballroom dancing medal with my husband and staying in my present job so long – that's quite an achievement these days.

**Good answer to question 19**   One big achievement was rearranging the back office, so that the existing team was able to deal with a lot more customer enquiries. The second big achievement was integrating several databases. I went on a special course to prepare for that, so that I was able to take over responsibility for the database conversion.

**Poor answer to question 20**   Yes, frequently, but my boss always stood in the way of change. One time, he did take up one of my suggestions, but then he claimed that it was his idea.

**Good answer to question 20**   Actually, I'm always looking for ways to improve. Sometimes I consult with colleagues first, but I have gone straight to my departmental manager with suggestions, too. The last suggestion I made was to set up standard form letters for order processing, so that we didn't have to type everything in every time.

21. How will you approach your new duties?

Your answer: _____

_____

_____

_____

_____

_____

_____

_____

_____

_____

22. How can you contribute to the organisation's success?

Your answer: _____

_____

_____

_____

_____

_____

_____

_____

_____

_____

**Poor answer to question 21**   With enthusiasm and interest.

**Good answer to question 21**   First of all, I need to see how things are done in your company and get to know out where to find the information I need. To that end, I'll talk to my new colleagues and my superiors. Then, I'll organise my work, so that, for example, urgent tasks can be dealt with quickly. And, finally, I'll deliver my results on schedule, as always.

**Poor answer to question 22**   I'll do my work as well and as efficiently as possible.

**Good answer to question 22**   Your organisation is one of the leading suppliers in the machine building sector and has an international profile. In the service department I'll be able to take on a client-side troubleshooting role. Because of my good command of English I could also undertake assignments abroad. More generally, I can offer my wide-ranging experience in machine tool manufacture.

23. In which areas are you a high achiever?

Your answer: _____

_____

_____

_____

_____

_____

_____

_____

_____

_____


24. How have you increased turnover in the past?

Your answer: _____

_____

_____

_____

_____

_____

_____

_____

_____

_____

**Poor answer to question 23**  I can do a lot of things. Lately, I haven't had much chance to prove myself, but I would still describe myself as very highly motivated.

**Good answer to question 23**  I respond well to demanding situations, and I'm able to work well with colleagues to achieve solutions. In one product series, for example, we had some initial quality problems. The call centre was very busy as a result. I got in touch directly with our colleagues in technical support to give clients concrete support in the event of operating problems with the new decoder.

**Poor answer to question 24**  I haven't. I'm not in sales.

**Good answer to question 24**  It might not be measurable directly, but by supporting the sales force as a marketing assistant I'm sure I've helped to increase turnover. I've had particular responsibility for event marketing, and I've increased awareness of our products in the marketplace.

25. What have you done to reduce costs in your present position?

Your answer: _____

_____

_____

_____

_____

_____

_____

_____

_____

_____

26. How can quality improvements be implemented in the company?

Your answer: _____

_____

_____

_____

_____

_____

_____

_____

_____

_____

**Poor answer to question 25**   Costing guidelines were always set by senior management in our firm. Often, they were unrealistic. For that reason, I don't think that drastic cost cutting measures are a good thing.

**Good answer to question 25**   Changes in the cost structure have been very important for my old employer in recent years. I was particularly involved in the introduction of new, less expensive packaging materials. I also contributed to minimising packaging bulk in order to reduce freight costs.

**Poor answer to question 26**   I think that can only be achieved through strict control systems.

**Good answer to question 26**   By every employee understanding that they must make a contribution to quality. Discussions in quality circles are certainly part of this, because they help employees to understand the concerns and needs of other departments. Fundamentally, I think it's important to remind ourselves all the time that quality isn't just a matter of chance, it's a goal that has to be worked towards by all concerned.

27. What are your ideas for increasing our customer base?

Your answer: _____

_____

_____

_____

_____

_____

_____

_____

_____

_____

28. In what way will we benefit from your joining the company?

Your answer: _____

_____

_____

_____

_____

_____

_____

_____

_____

_____

**Poor answer to question 27**   You could introduce new products or massively increase your advertising.

**Good answer to question 27**   These days money is tight. I've had good experiences with innovative but cost-effective measures. For example, I increased sales at trade fairs and made sure that we reach more small businesses with special product packages.

**Poor answer to question 28**   You'll be getting a highly motivated employee.

**Good answer to question 28**   You'll be getting a new employee who is familiar with your day-to-day business and can do useful work from day one. Also, I know from past experience that I can motivate my co-workers, and that will be important when the new branch office is set up.

# 5

# Are you customer-oriented?

Die Bedeutung einer klar auf den Kunden ausgerichteten Geschäftsstrategie hat in den letzten Jahren immer weiter zugenommen. Insbesondere Bewerber aus den Bereichen Verkauf, Vertrieb, Marketing, Service und Beratung werden deshalb mit ausführlichen Fragen zu ihrer Kundenorientierung rechnen müssen.

## Hintergrund

Vor allem in engen und gesättigten Märkten verkaufen sich Produkte oder Dienstleistungen nicht von allein. Ihre Vorteile und Besonderheiten müssen potenziellen Kunden deshalb in beratungsintensiven Gesprächen geschickt vermittelt werden können. Da der Kontakt zwischen Kunde und Firma über die Schnittstellen Verkauf und Marketing, aber auch über den Service stattfindet, möchten die Firmen von Bewerbern um diese Stellen anhand anschaulicher Beispiele erfahren, wie sie vorgehen, um neue Kunden zu gewinnen und bestehende Kunden an die Firma zu binden.

## Typische Fehler

Allgemein gehaltene Lippenbekenntnisse zur Bedeutung der Kundenorientierung im Zeitalter der Austauschbarkeit von Produkten oder Dienstleistungen helfen hier nicht weiter. Genauso gefährlich sind in Anklageform vorgetragene Monologe zur Servicewüste Deutschland. Sie werden Ihre Gesprächspartner auch nicht von Ihrer eigenen Kundenorientierung überzeugen können, wenn Sie Missstände bei Ihrem aktuellen Arbeitgeber ausführlich auflisten oder die Fehler Ihrer Kollegen kritisieren. Genauso problematisch ist es, mit der inneren Einstellung *I can sell anything!* aufzutreten. Jede Firma hängt an ihren speziellen Produkten oder Dienstleistungen, und wer nicht glaubwürdig vermitteln kann, dass er sich mit den Besonderheiten der Angebotspalette gründlich auseinandergesetzt hat, wird im Vorstellungsgespräch Schiffbruch erleiden.

### Negativbeispiel

Die Frage *What does ›customer focus‹ suggest to you?* ist eigentlich eine tolle Chance, um sich als Bewerber ins richtige Licht zu setzen. Leider gibt es dennoch Kandidaten, die diese Steilvorlage beispielsweise so vergeben: *It's a slogan like so many others these days. The main thing is to do good work, then the customer will be happy. The customer doesn't really know what he wants until you make him an attractive offer.*

### Kommentar zum Negativbeispiel

Mit dieser Antwort gibt der Bewerber zu verstehen, dass er den Kunden eigentlich für einen Störenfried hält. Er möchte sich lieber in seinem Arbeitsbereich einigeln und möglichst wenig über den eigenen Tellerrand schauen. Auch seine Einstellung zum Absatz von Produkten ist sehr fragwürdig: Anscheinend glaubt er tatsächlich, dass sich der Kunde alles aufs Auge drücken lässt, wenn man es ihm nur geschickt genug anbietet. Dass dabei keine langfristige Kundenbindung entstehen kann, ist diesem Bewerber anscheinend egal.

## Antwort-Strategie

Die Erfahrung zeigt, dass berufserfahrene Bewerber aus den Bereichen Verkauf, Marketing und Service über einen reichen Fundus an Beispielen für gelebte Kundenorientierung verfügen. Überlegen Sie sich also vor dem Gespräch, welche Beispiele aus Ihrer Berufspraxis am besten zu der ausgeschriebenen Stelle passen. Stellen Sie sich als jemand dar, der immer wieder aufs Neue Freude daran hat, Kunden von der Qualität seiner Produkte oder Dienstleistungen zu überzeugen. Zeigen Sie auch auf, dass Sie sich mit dem Erreichten niemals zufriedengeben, sondern permanent an einer Verbesserung der Stellung am Markt arbeiten.

## Positivbeispiel

Dass man die recht offene Frage *What does ›customer focus‹ suggest to you?* besser nutzen kann, um sich in ein gutes Licht zu rücken, zeigt diese gelungene Antwort: *It suggests to me a lot of opportunities for companies. On your website I saw that you offer tailored solutions to customers. I think that's a very promising approach. In marketing I've always focussed on developing offers which are tailored to the target group. To do that, I brought my colleagues from development, production and customer service to the table. By using interdepartmental project teams to improve our customer orientation we were able to focus the work of each department on the specific needs of the customer.*

## Kommentar zum Positivbeispiel

Mit seiner Antwort sammelt der Bewerber viele Pluspunkte. Er verdeutlicht, dass er Kundenorientierung nicht bloß für eine vorübergehende Mode hält, sondern sie in seiner Arbeit lebt. Zudem stellt er heraus, dass er sich mit den Angeboten der neuen Firma intensiv beschäftigt hat. Diese Vorgehensweise wird ihn weiterbringen und die Personalverantwortlichen überzeugen, denn schließlich geht es darum, die eigenen Kenntnisse und Erfahrungen so zu präsentieren, dass der Nutzen für den neuen Arbeitgeber deutlich wird. Deshalb stellt der Bewerber heraus, dass er jetzt schon die gleichen Maßstäbe an seine Arbeit anlegt, die auch die neue Firma gerne verwirklicht sehen möchte.

29. What contribution can you make in your field of work to help us win more customers?

Your answer: _____

_____

_____

_____

_____

_____

_____

_____

_____

_____

30. In your view, what do customers value about our products / services?

Your answer: _____

_____

_____

_____

_____

_____

_____

_____

_____

_____

**Poor answer to question 29**  I think I would advocate price reductions.

**Good answer to question 29**  In production it's very important that no products leave the hall with defects of any kind. In previous jobs I've been involved in quality assurance groups. So I know that we in production have to report back if manufacturing stages become so complicated that errors can occur. If we in production take care, the quality and reliability of our products can be improved – and then more customers will want them.

**Poor answer to question 30**  Well, people can't do their own tax returns these days, it's all too complicated. People need a tax adviser.

**Good answer to question 30**  That they feel they're thoroughly taken care of. You offer a comprehensive service in your tax consultancy. Not just taxation advice, but also bookkeeping, company start-ups, help with inheritance issues and even property management. Clients get a complete package.

31. In your view, what do customers dislike about our products/
    services?

Your answer: _____

_____

_____

_____

_____

_____

_____

_____

_____


32. What sort of experience with customers have you had in your
    present job?

Your answer: _____

_____

_____

_____

_____

_____

_____

_____

_____

**Poor answer to question 31**  It's noticeable that your products are quite expensive. I'm sure that some customers must wonder whether these prices are really justified.

**Good answer to question 31**  With brand-name goods, some customers are going to ask themselves how much they are paying just for the brand. I think it's important to show people that the product quality really is better. You could point out the many good test results you've had. Leaving the hard facts aside, though, it's important to give the customer the feeling that they are getting a superior product.

**Poor answer to question 32**  Good and bad, according to the products I had to sell.

**Good answer to question 32**  Contact with the customer is very important to me. I've always looked at customer complaints as a chance to improve the product. And positive feedback from customers gives me added motivation. The most important thing is that customers feel that they are being taken seriously and are being offered a product which matches their needs.

33. Imagine a friend tells you about a negative story in the press about our company. How would you react?

Your answer: _____

_____

_____

_____

_____

_____

_____

_____

_____

_____

34. Is customer focus important in your job?

Your answer: _____

_____

_____

_____

_____

_____

_____

_____

_____

_____

**Poor answer to question 33**   The press is full of exaggerations. There are plenty of other companies which have problems.

**Good answer to question 33**   I would try to put things in the right perspective. In the press, things often get blown out of proportion or presented incorrectly. I would definitely point to the good reputation that the company has built up over the years and say something positive about the company.

**Poor answer to question 34**   I don't have any direct contact with customers, so I don't think it's very important.

**Good answer to question 34**   Customer focus is always important. Even though I don't have any direct contact with customers, it's essential to keep the customer in mind. After all, the other departments, which have to use our output, are internal customers in a way. I always make an effort to produce work which is genuinely useful to others.

35. Which is more important: good marketing or good products?

Your answer: _____

_____

_____

_____

_____

_____

_____

_____

_____

_____


36. What can be done to strengthen employees' customer focus?

Your answer: _____

_____

_____

_____

_____

_____

_____

_____

_____

_____

**Poor answer to question 35**   A good product will always find its way to the customer.

**Good answer to question 35**   Good marketing and good products should go hand-in-hand. It's no use having a good product if nobody knows about it. With the complex technical products that you produce, their full capabilities aren't apparent at first glance. For that reason, marketing is important to guide the customer and provide information. We technicians supply the good product which matches the marketing message.

**Poor answer to question 36**   If you don't see the writing on the wall, you're going to fail. Some people have to learn their lessons the hard way.

**Good answer to question 36**   In the end, every job depends on satisfied customers. That's why I think that every employee needs to recognise the value of their contribution to the success of the company. Communication within the company is important, so that feedback from sales and customer service reaches the people in product development and administration. Interdepartmental project groups can help with that.

37. A customer complains to you about a faulty product from our company. How do you react?

Your answer: _____

_____

_____

_____

_____

_____

_____

_____

_____

_____

38. How can we win long-term customer loyalty?

Your answer: _____

_____

_____

_____

_____

_____

_____

_____

_____

_____

**Poor answer to question 37**  I'd pass the message on to the responsible person in the company for them to take care of.

**Good answer to question 37**  I would take the complaint seriously and find out what exactly the customer thinks is wrong with the product. Then I would offer a solution. It might be a repair or a replacement product. The important thing is that the customer buys from us again next time, in spite of the complaint.

**Poor answer to question 38**  A difficult question. These days, customers come and go faster than suppliers.

**Good answer to question 38**  In my experience, there's a lot that can be done to encourage customer loyalty. Competent advice during the sale can lead the customer to come back next time. Also, I've always found a customer database to be an effective tool, so that follow-ups and mail shots can be used to introduce new products.

39.  Are you experienced in talking to customers?

Your answer: _____

_____

_____

_____

_____

_____

_____

_____

_____

_____

40.  What do you think of the sentence ›You can't learn to sell, it's either in your blood or it isn't‹?

Your answer: _____

_____

_____

_____

_____

_____

_____

_____

_____

_____

_____

**Poor answer to question 39**  Yes, of course. That's my job, after all.

**Good answer to question 39**  Yes, I enjoy advising customers. I like finding out what they want. A lot of customers don't know exactly what they are looking for when they walk in the door. I'm glad to be able to help them and recommend a particular product.

**Poor answer to question 40**  There is something in that. All the same, there's no harm in learning a new trick or two.

**Good answer to question 40**  You need to enjoy dealing with customers, otherwise you shouldn't be in sales. On the other hand, a lot of things have to be learned – talent alone is no replacement for a comprehensive knowledge of the product. After all, you don't just want to make a sale, you want to give competent advice, too.

41. What can we do to maintain our good reputation with our customers into the future?

Your answer: _____

_____

_____

_____

_____

_____

_____

_____

_____

42. How up-to-date is the saying ›The customer is king.‹ in your view?

Your answer: _____

_____

_____

_____

_____

_____

_____

_____

_____

**Poor answer to question 41**   Give me the job! Then your reputation is in safe hands.

**Good answer to question 41**   The good reputation of your products was the deciding factor in my decision to apply for this job. If you continue to value quality, innovation and customer-friendliness this highly, you will continue to have a good reputation in the market.

**Poor answer to question 42**   There are customers who spend 10 euros and think they can take liberties. I'm not sure that we really need to give every customer special treatment.

**Good answer to question 42**   I think this saying is very up-to-date. There are a lot more vendors than there used to be. The customer is the one who decides where to buy, so every customer should be highly valued by all employees of the company.

43. If you were one of our customers: what would be important to you?

Your answer: _____

_____

_____

_____

_____

_____

_____

_____

_____

44. Which new sales channels could be leveraged to reach more customers?

Your answer: _____

_____

_____

_____

_____

_____

_____

_____

_____

**Poor answer to question 43**   The important factors for me would be price, quality and service.

**Good answer to question 43**   For me, it would be important to get a good product at the right price. Competent technical advice would make the buying decision easier, of course. And the guarantee of a good service relationship would be very important to me, too.

**Poor answer to question 44**   I'll need to think about that. I'm sure I can come up with something. Perhaps a creative session with colleagues would be a good step, too, to generate new ideas.

**Good answer to question 44**   You are already leveraging the usual sales channels to the maximum. One possibility that I can see would be the shop-in-shop concept. As a manufacturer of quality clothing accessories you could position yourselves as a vendor in chain stores and give your brand name greater reach in that way.

# 6

# How good are your PC skills?

Neben speziellen Fachkenntnissen, Branchenwissen und persönlichen Fähigkeiten des Bewerbers werden in sehr vielen Stellenanzeigen auch Fremdsprachen- und PC-Kenntnisse eingefordert. In zahlreichen Berufen gelten besonders PC-Kenntnisse mittlerweile als Standardvoraussetzung. Deshalb müssen Sie Ihre Fähigkeiten in diesen Bereichen plausibel belegen können.

## Hintergrund

Um unliebsamen Überraschungen – nämlich fehlenden Sprach- oder EDV-Kenntnissen – im späteren Berufsalltag vorzubeugen, wird im Vorstellungsgespräch auch überprüft, inwiefern der Bewerber entsprechende Vorgaben der Firmenseite erfüllt. Erstaunlicherweise bereitet sich so mancher Bewerber auf Fragen zu Sprach- und PC-Kenntnissen häufig schlechter vor als auf gängige Fragen zur Selbstmotivation, zur Firma, zur Leistungsbereitschaft oder zum Führungsverhalten. Das kann jedoch zu einem Bumerang für den Bewerber werden, denn oft gelten diese Kenntnisse als wichtige Grundvoraussetzung.

## Typische Fehler

Auch bei Fragen zu Fremdsprachen- oder PC-Kenntnissen gilt: Wer in seine Antworten keine konkreten Beispiele einfließen lässt, wird es im Vorstellungsgespräch sehr schwer haben. So sollten Sprachkenntnisse nicht nur aufgezählt werden, sondern es sollten vielmehr Gelegenheiten genannt werden, in denen sie erfolgreich eingesetzt wurden – beispielsweise im Umgang mit Kunden, bei Präsentationen, in Meetings oder auf Kongressen. Und auch vorhandene PC-Kenntnisse sollten nicht bloß heruntergeleiert, sondern in Anwendungssituationen erläutert werden.

### Negativbeispiel

Auf die Frage *What are your computer skills like?* sollten Bewerber konkret antworten. Ausführungen wie die folgende wären hingegen zu knapp: *Oh, I'm familiar with the usual software that's used in the back office. You can't get anywhere without IT these days. So my computer skills are good.*

### Kommentar zum Negativbeispiel

Die Antwort der Bewerberin ist deutlich zu knapp. Wahrscheinlich ist sie in die »Selbstverständlichkeitsfalle« getappt. Es passiert häufiger, dass Bewerber, die tagtäglich mit bestimmten Softwareprogrammen umgehen, ihre Kenntnisse für selbstverständlich und nicht weiter erwähnenswert halten. Für Außenstehende wie Personalverantwortliche und andere Firmenvertreter sieht die Sache aber anders aus: Sie möchten ganz konkret erfahren, welche Programme die Bewerberin beherrscht.

### Antwort-Strategie

Wenn es um Ihre PC-Kenntnisse geht, sollten Sie deutlich machen, dass Sie sich auch in der Vergangenheit immer wieder in neue PC-

Programme eingearbeitet haben. Machen Sie die von Firmen geforderte Bereitschaft zum lebenslangen Lernen konkret: Liefern Sie Beispiele dafür, wie Sie Ihre EDV-Kenntnisse in speziellen Kursen, autodidaktisch oder durch eine interne Einarbeitung immer wieder erweitert und ausgebaut haben.

## Positivbeispiel

Mit etwas mehr Liebe zum Detail kann die Frage *What are your computer skills like?* viel besser beantwortet werden: *In the back office I work with the Microsoft Office package on a daily basis. For correspondence I use Word, I use Excel to present statistics, and I'm also experienced in creating presentations in PowerPoint. I'm a proficient e-mail user, and I can use the internet for research.*

## Kommentar zum Positivbeispiel

Hier bringt die Bewerberin ihre EDV-Kenntnisse richtig ins Spiel. Sie wird konkret und erläutert, wie sie typische Programme bei der täglichen Arbeit einsetzt. Diese Bewerberin ist auf der Höhe der Zeit. Man kauft ihr ohne weiteres ab, dass sie sich auch künftig in Programm-Updates oder neue Programme einarbeiten wird. Die Antwort ist eine glaubwürdige Darstellung ihrer PC-Praxis.

45. Can you conduct conversations with customers in English?

Your answer: _____

_____

_____

_____

_____

_____

_____

_____

_____

_____

46. When did you last read a professional article in English?

Your answer: _____

_____

_____

_____

_____

_____

_____

_____

_____

_____

**Poor answer to question 45**   Yes!

**Good answer to question 45**   Yes, I could. In my last job I dealt with international customers. As suppliers we had a worldwide customer base, so English was our main language of business.

**Poor answer to question 46**   Never. I sometimes look at English song lyrics, but apart from that ...

**Good answer to question 46**   A few days ago. I was researching a technical issue on the internet, and of course a lot of the relevant pages were in English.

47. Are you confident that you could conduct negotiations in English?

Your answer: _____

_____

_____

_____

_____

_____

_____

_____

_____

_____


48. What have you done in the last two years to keep your English up-to-date?

Your answer: _____

_____

_____

_____

_____

_____

_____

_____

_____

_____

**Poor answer to question 47**   I'm not sure. If I had an interpreter with me, it wouldn't be a problem.

**Good answer to question 47**   I use English to communicate with colleagues and customers. When it comes to specific details of a contract, I'd prefer to be cautious though. Otherwise, I can conduct negotiations in English and have done so in the past.

**Poor answer to question 48**   Learning a language is like learning to ride a bike. You never really forget.

**Good answer to question 48**   I read online articles from English-speaking newspapers at least once a week. And in my work I'm increasing my technical vocabulary all the time.

49. Which applications do you use for which tasks?

Your answer: _____

_____

_____

_____

_____

_____

_____

_____

_____

_____

_____

50. How did you acquire your software knowledge?

Your answer: _____

_____

_____

_____

_____

_____

_____

_____

_____

_____

_____

**Poor answer to question 49**   The ones that are appropriate – a word-processing application for letters and other suitable software.

**Good answer to question 49**   I work with Microsoft Office on a daily basis – Word for correspondence, Excel for statistics and Power-Point for presentations. On top of that, I also use specialist measuring and calculating software.

**Poor answer to question 50**   As I went along, by trial and error. I would have liked more support from my company. I'm sure I could do a lot more with the software if only I knew how.

**Good answer to question 50**   I taught myself to use Word with the help of tutoring CDs in my own time. The same goes for Power-Point. To learn Excel, I did an advanced course at evening school. To learn my company's specialist software, I did in-house training.

51. How do you go about learning a new software package?

Your answer: _____

_____

_____

_____

_____

_____

_____

_____

_____

_____

52. Which applications would you like to learn in more depth?

Your answer: _____

_____

_____

_____

_____

_____

_____

_____

_____

_____

**Poor answer to question 51**   I try things out and ask colleagues who know how to use it.

**Good answer to question 51**   With standard software you can learn to use some functions if you are familiar with other applications. Otherwise, I've always found tutoring programs useful. If I'm still not quite clear about particular functions, I use the application's own help system to find out exactly what I need to know. I also ask colleagues for tips.

**Poor answer to question 52**   Which applications would you like me to learn?

**Good answer to question 52**   I'm interested in any applications that can help me in my work. I use trade journals to get an overview of the new products on the market. In particular I'd be interested in new route planning software which you can quickly customise with your customers' details.

# 7

# What do you know about our company?

Wunschkandidaten können im Vorstellungsgespräch vermitteln, dass sie in zweifacher Weise in das Unternehmen passen: Sie machen deutlich, dass sie sowohl auf die neue Stelle als auch in die neue Firma passen. Um das zu überprüfen, stellen die Entscheider auf der Firmenseite nicht nur Fragen zum neuen Arbeitsplatz, sondern auch zur geschäftlichen Entwicklung der Firma. Für Sie als Bewerber ist es deshalb wichtig, sich vorab ausführlich über die Firma zu informieren. Sammeln Sie Informationen auf der Firmenhomepage oder in Zeitungen, oder lassen Sie sich direkt vom Unternehmen Infomaterialien schicken – diese können Sie in vielen Fällen zum Beispiel in der PR-Abteilung der Firma anfragen.

## Hintergrund

Die Art und Weise, wie Bewerber Fragen zur Firma beantworten, ist für Personalverantwortliche in mehrfacher Hinsicht aufschlussreich: Zum einen lässt sich daran erkennen, wie ernsthaft die Bewerbung gemeint ist, da sich interessierte Bewerber auf diese Fragen üblicherweise gut vorbereiten. Zum anderen werden die Antworten als Arbeitsprobe für die Firma gedeutet. Man will erfahren, ob der Bewerber die unausgesprochene Aufgabe *Bereiten Sie das Vorstellungsgespräch gründlich vor* erkannt und ernst genommen hat.

## Typische Fehler

Bewerber, die allgemein zugängliche Kennzahlen der neuen Firma nicht parat haben, sorgen für Missstimmung. Wer Fragen nach der Anzahl der Beschäftigten, nach Umsätzen und Gewinnen der vergangenen Jahre oder nach weiteren Firmenstandorten im In- und Ausland nicht beantworten kann, disqualifiziert sich schon selbst. Gleiches gilt für Fragen zu den wichtigsten Produkten beziehungsweise Dienstleistungen der Firma. Es darf auf keinen Fall der Eindruck entstehen, dass Ihnen eigentlich egal ist, in welcher Firma Sie arbeiten.

## Negativbeispiel

Eine typische Frage in diesem Fragenblock ist: *Do you know how we make our money?* Antworten wie die folgende bringen Bewerber allerdings nicht weiter: *Well, as a call centre operator I expect you make your money from customer calls. You're a big name in this business. As companies are outsourcing more and more of their customer service, I expect you're very busy. Every company plays to its strengths, and call centres are the experts in telecommunications.*

## Kommentar zum Negativbeispiel

Diese Antwort kann nicht überzeugen, passt sie doch zu jedem Call-Center-Betreiber gleich gut – genauer gesagt: gleich schlecht. Jeder, auch ein Unternehmensvertreter, möchte als einzigartig wahrgenommen werden. Wer dem Firmenvertreter im Gespräch also das Gefühl gibt, dass sein Unternehmen eigentlich austauschbar ist, stört die Gesprächsatmosphäre nachhaltig. Es wird auch unangenehm aufstoßen, dass der Bewerber sich nicht genauer über den möglichen Arbeitgeber informiert hat. Eine lieblose beziehungsweise fehlende Recherche deutet darauf hin, dass der Kandidat auch beim Arbeiten öfter unvorbereitet agiert.

## Antwort-Strategie

Mit dem gezielten Einsatz des Internets lassen sich ohne großen Aufwand die wichtigsten Informationen über den neuen Arbeitgeber recherchieren. Gehen Sie also auf die Homepage der Firma, und geben Sie den Firmennamen in Suchmaschinen ein. Oder lassen Sie sich bei größeren Unternehmen Infomaterial direkt von der Firma schicken. Betonen Sie dann in Ihren Antworten, dass Sie sich vor dem Gespräch gründlich über die Firma informiert haben. Die Vorbereitung des Vorstellungsgesprächs sollten Sie unbedingt ernst nehmen, betrachten Sie sie als eine weitere Arbeitsprobe – genau wie die vorherige schriftliche Bewerbung auch. Besonders gut macht es sich zudem, wenn Sie wichtige Mitbewerber kennen und darstellen können, welche Chancen und Risiken Sie für die zukünftigen Entwicklungen der Branche sehen.

## Positivbeispiel

Mit etwas Recherche lässt sich die Frage *Do you know how we make our money?* viel besser beantworten, nämlich so: *Yes. Your corporate website was my main source of information. Your biggest clients include Telecom International, United Telecom and TelecomPlus. You are the leading supplier of call centre solutions for telecoms companies. You don't just offer a fault hotline, you provide a full service package. You ensure that faults are dealt with actively, by using the call centre to narrow down the problem before passing it on to technical subcontractors. On top of that you deal with upgrades of existing contracts, for example broadband for customers with dial-up or the Home Call option for mobile phone contracts.*

## Kommentar zum Positivbeispiel

Mit etwas Vorbereitung lassen sich wie in dieser gelungenen Antwort viele Pluspunkte bei den Firmenvertretern sammeln. Diesem Bewerber wird im weiteren Verlauf des Gespräches mehr Vertrauen und Aufmerksamkeit entgegengebracht werden als anderen Kandidaten. Schließlich kann man aus seiner Antwort entnehmen, dass er sich aktiv mit seinem zukünftigen Arbeitgeber beschäftigt hat: Er kann die Besonderheiten im Service nennen und das Unternehmen somit von Konkurrenten abgrenzen. Dies unterstreicht die Ernsthaftigkeit seiner Bewerbungsabsichten und verschafft ihm einen Sympathiebonus.

53. What is the central problem in our industry?

Your answer: _____

_____

_____

_____

_____

_____

_____

_____

_____

_____

_____

54. Are you familiar with our corporate website?

Your answer: _____

_____

_____

_____

_____

_____

_____

_____

_____

_____

_____

_____

**Poor answer to question 53**   Things aren't going well in any sector of industry at the moment. Times are hard, so I expect you're under pressure, too.

**Good answer to question 53**   In my view, the central problem is low profit margins. I think, direct sales would be one way of improving profitability. I had some success in that area with my last employer.

**Poor answer to question 54**   Yes, I had a look at it.

**Good answer to question 54**   I prepared thoroughly for this interview, so, of course, I had a good look at your website. I liked the structure and the clarity. You can find what you are looking for and navigate easily between the different sections.

55. Which of our locations would you most like to work at?

Your answer: _____

_____

_____

_____

_____

_____

_____

_____

_____

_____


56. Do you know how many employees we have?

Your answer: _____

_____

_____

_____

_____

_____

_____

_____

_____

_____

_____

**Poor answer to question 55**   Actually, I wouldn't want to work here in Stuttgart. I'd most like to work in North Rhine-Westphalia. I know lots of people there.

**Good answer to question 55**   I chose to apply for the advertised position at your Stuttgart office. On the other hand, I see on your website that you have offices elsewhere. If required, I would be prepared to work at one of them for a time.

**Poor answer to question 56**   I think about 400. Or was it 1,400? I read somewhere that it was even more. But I'm not sure of the exact number.

**Good answer to question 56**   Here in Stuttgart you have over 400 employees, nationwide there are just under 1,500. In the whole of Europe you have about 2,000 employees.

57. Do you know how many branches we have?

Your answer: _____

_____

_____

_____

_____

_____

_____

_____

_____

_____

58. Where did you get your information about our company?

Your answer: _____

_____

_____

_____

_____

_____

_____

_____

_____

_____

**Poor answer to question 57**   No, I'm afraid not.

**Good answer to question 57**   I know that you have 12 branches in Germany, with your headquarters in Essen.


**Poor answer to question 58**   The job advert had some of the main points, and I read an article about you in the newspaper a while ago.

**Good answer to question 58**   I found out as much as I could about you. First of all, I looked at your website. Then, I used a search engine to find more information about specific products and campaigns. Also, your colleagues in the PR department were kind enough to send me further information, about your corporate guidelines, for example.

59. What impression do you have of our company?

Your answer: _____

_____

_____

_____

_____

_____

_____

_____

_____

_____

_____


60. Where did you hear of our company?

Your answer: _____

_____

_____

_____

_____

_____

_____

_____

_____

_____

_____

**Poor answer to question 59**   A very good one so far. But I'll be working in the field, in any case.

**Good answer to question 59**   A very professional impression. There's an efficient, friendly atmosphere here. If I were a prospective customer, I would feel I was in good hands.


**Poor answer to question 60**   From the job advert. That was the first time I heard of you.

**Good answer to question 60**   I've known of your company for several years. My first contact with you was at a trade fair. After that I often came across articles about you. I've been impressed time and again by your company's spirit of innovation.

61. What attracts you to our industry?

Your answer: _____

_____

_____

_____

_____

_____

_____

_____

_____

_____


62. Are you familiar with our competitors?

Your answer: _____

_____

_____

_____

_____

_____

_____

_____

_____

_____

_____

**Poor answer to question 61**   I tried to switch over to another industry, but I just don't seem to be able to get away from the car industry, and, actually, it's quite interesting.

**Good answer to question 61**   I'm particularly attracted by the pace of technical development in the industry. As a sales executive I'm aware of the technical innovations in each new model. It's always a challenge to present that to the customer in a readily understandable way.

**Poor answer to question 62**   At least with one – the one I work for.

**Good answer to question 62**   I've been in the industry for several years, and it's part of the job to have an overview of the other suppliers in the market. Your most important competitors would include Müller GmbH, Schmidt AG and Meyer GmbH & Co. KG. But overseas suppliers like Trading Corp. are soon going to have a stronger presence in Germany, too.

63. Are you familiar with our products/services?

Your answer: _____

_____

_____

_____

_____

_____

_____

_____

_____

_____


64. What do you think could be improved about our business?

Your answer: _____

_____

_____

_____

_____

_____

_____

_____

_____

_____

**Poor answer to question 63**   I know some of them, and they're good products, too.

**Good answer to question 63**   You're particularly strong in the area of custom lighting systems. Your natural lighting concept for open-plan offices is really interesting, too. Then, there's your lighting systems for concert halls and multi-purpose venues.


**Poor answer to question 64**   The first thing is that you need to review your company policy towards suppliers. I know how much pressure you put on them; that has to result in poor quality. Secondly, you have too many people in the office. Thirdly, your field sales operatives aren't very motivated. And I'm sure I can think of other things that need urgent attention, too.

**Good answer to question 64**   Your products have a good position in the market. It might be interesting to think about synergy between the different product lines. It's become a big trend in the industry lately to make greater use of customer service teams to foster customer loyalty and increase sales. I think that you could refine a few details to build on your strong market presence.

# 8

# How do you cope with change?

Der Veränderungsdruck, dem die Firmen eigentlich schon immer ausgesetzt waren, hat in den vergangenen Jahren enorm zugenommen. Gründe für die notwendigen Veränderungen sind unter anderem die schwierige wirtschaftliche Lage und die zunehmende Konkurrenz aus dem Ausland. Im Vorstellungsgespräch möchte man herausbekommen, ob Sie diesem Druck auf Dauer standhalten können.

## Hintergrund

Restrukturierungen, Kostensenkungsprogramme, Abteilungsumgestaltungen oder Bereichszusammenlegungen finden in Firmen immer häufiger statt. Es nützt aber nichts, wenn die Firmenspitze oder externe Unternehmensberater »von oben herab« neue Konzepte entwickeln und implementieren. Diese Konzepte entfalten erst dann ihre Wirkung, wenn sie von allen Beteiligten ernst genommen und umgesetzt werden. Deshalb sind Firmen daran interessiert, zu erfahren, wie Sie in der Vergangenheit mit Veränderungen im Berufsalltag umgegangen sind.

## Typische Fehler

Für die meisten Bewerberinnen und Bewerber sind als Arbeitnehmer die genannten Veränderungen mit gravierenden Einschnitten wie Lohnkürzungen, Etatstreichungen oder oft auch Arbeitsplatz-abbau verbunden. Daher kann es schnell passieren, dass in den Antworten auf die Fragen zur Veränderungsbereitschaft die Emotionen durchschlagen und eine pauschale Managerkritik oder Arbeitgeberschelte betrieben wird. So eine Reaktion wirkt sich im Vorstellungsgespräch mit dem neuen Arbeitgeber natürlich kontraproduktiv aus.

### Negativbeispiel

Die Veränderungsbereitschaft von Bewerberinnen und Bewerbern lässt sich mit der Frage *What aspects of your work have changed in recent years?* überprüfen. Auf keinen Fall dürfte die Antwort so ausfallen: *Employers always try to put you under pressure, and that discourages any sort of commitment on the part of the employee. Also, we have to do everything on the computer these days, and I think that's pointless. It isn't necessary to record absolutely everything, that turns you into just another office drone in the long run.*

### Kommentar zum Negativbeispiel

Sicherlich gibt es immer wieder mal etwas an bestehenden Arbeitsabläufen zu kritisieren. Allerdings ist das Vorstellungsgespräch ein völlig ungeeigneter Ort, um diese Kritik zu äußern. Mit seiner Antwort wirft der Bewerber dem Personalverantwortlichen indirekt an den Kopf, dass er alle Vorgesetzten für bürokratische Monster hält. Er scheint zudem nicht sonderlich viel davon zu halten, sich an neue Entwicklungen anzupassen. Durch seine Pauschalkritik manövriert er sich jedoch ins Abseits.

## Antwort-Strategie

Machen Sie klar, dass Sie Veränderungen grundsätzlich weniger als Bedrohung, sondern vielmehr als Chance und Herausforderung sehen. Betonen Sie Ihre Fähigkeit, sich flexibel auf veränderte Anforderungen einzustellen. Liefern Sie Beispiele dafür, wie Sie in Zeiten knapper Kassen und dünner Personaldecken mit den Aufgaben in Ihrem Arbeitsbereich dennoch zurechtgekommen sind. Sehr überzeugend sind auch Beispiele dafür, wie Sie Veränderungen mithilfe kreativer – sprich: kostenneutraler – Lösungen erfolgreich in den Griff bekommen haben.

## Positivbeispiel

Damit Sie die Frage *What aspects of your work have changed in recent years?* besser als in dem vorherigen Negativbeispiel beantworten können, sollten Sie sich an der folgenden Antwort orientieren: *Many things have changed. For example, there's the increased use of IT. You can get key data more quickly than before, and better targeted to your requirements. I've always seen that as a great opportunity and have been quick to learn the relevant software. It has also become more important to consult with other departments, so regular interdepartmental meetings are part of my work these days. It might take more time and effort, but it's worth it to get all participants pulling in the same direction.*

## Kommentar zum Positivbeispiel

Die Forderung der Firmen nach flexiblen Mitarbeitern greift der Bewerber in dieser Antwort gekonnt auf. Auch seine Erfahrungen mit Veränderungen werden sicherlich nicht ausschließlich positiv sein, aber für das Vorstellungsgespräch trifft er die richtige Auswahl an geeigneten Beispielen. Er beschränkt sich nicht nur darauf, die Veränderungen am Arbeitsplatz aufzuzählen, sondern liefert auch Informationen darüber, wie er mit dem Veränderungsdruck umgegangen ist – und diese Beispiele wirken sehr überzeugend. Diesem Bewerber traut man ohne weiteres zu, dass er es schafft, sich auf die Anforderungen im neuen Job einzustellen. Damit ist er einen wesentlichen Schritt auf dem Weg zur Einstellung weitergekommen.

65. Have you developed professionally in recent years?

Your answer: _____

_____

_____

_____

_____

_____

_____

_____

_____

_____


66. Name two major changes you experienced in your last job and
    tell us how you coped with them.

Your answer: _____

_____

_____

_____

_____

_____

_____

_____

_____

_____

**Poor answer to question 65**   Yes, I hope so.

**Good answer to question 65**   Definitely. In my area of expertise I always keep up with new developments. The internet is a great source of up-to-date information. I've ›grown into‹ more difficult assignments, too. There are also the special assignments I've taken on, which have given me a closer relationship with colleagues from other departments.

**Poor answer to question 66**   Once, we had a change of manager, which was actually a good thing for the department. And then, there was the move to the new premises. I made sure I got a good office this time.

**Good answer to question 66**   I've been involved in several changes in production. The biggest challenge was the introduction of the three-shift system. It wasn't easy, because I had to make changes in my private life. Then, we had a lot of managerial changes within a short period of time. I always got on well with the new managers, but of course you have to get used to the new boss' way of doing things. There was one young manager who I gave a lot of support to, because he had come from outside and didn't have many contacts within the company.

67. Have you ever experienced budget cuts in your own workplace? How did you cope with them?

Your answer: _____

_____

_____

_____

_____

_____

_____

_____

_____

_____

68. Give me two examples of your professional flexibility.

Your answer: _____

_____

_____

_____

_____

_____

_____

_____

_____

_____

**Poor answer to question 67**   Budget cuts are a fact of life, even if they do cause a lot of disruption.

**Good answer to question 67**   It isn't easy when your budget is cut time after time. In my department we lost two out of ten jobs. The remaining colleagues had to divide up the work between them. Of course, that meant more work for everyone, but the workload was still manageable. Our advertising budget was cut as well. Together with the rest of the team I made sure that the remaining budget was only used for selected advertising channels with a high attention value.

**Poor answer to question 68**   I had to relocate for my last employer, and I even had to cancel my leave once.

**Good answer to question 68**   I've often covered for colleagues, once for an extended period. And I've taught myself to use new software more than once.

69. How do you help colleagues to adapt to changed work processes?

Your answer: _____

_____

_____

_____

_____

_____

_____

_____

_____

_____

70. Describe a professional experience which has had a decisive influence on you.

Your answer: _____

_____

_____

_____

_____

_____

_____

_____

_____

_____

**Poor answer to question 69**  It depends. I don't mind giving nice colleagues a tip sometimes. Otherwise, they can look after themselves.

**Good answer to question 69**  I talk to them about how I approached the new work processes. When it comes to technical matters, of course, I'm always ready to give my colleagues advice. It's always best to consult each other, that way everything runs more smoothly.

**Poor answer to question 70**  There was the bankruptcy of the company where I did my training. In that sort of situation you realise that even your best efforts can be in vain.

**Good answer to question 70**  My first appointment to a project group. That's where I learned how all the processes in the company fit closely together. Since then I've looked beyond the needs of my own department a lot more.

71. What would cause you to look for a new employer?

Your answer: _____

_____

_____

_____

_____

_____

_____

_____

_____

_____

72. Can you adapt to new situations easily?

Your answer: _____

_____

_____

_____

_____

_____

_____

_____

_____

_____

**Poor answer to question 71**   The warning signs are usually clear enough. It's better to make the move sooner rather than later.

**Good answer to question 71**   Being no longer able to make a worthwhile contribution. I want to use my professional experience to achieve results. If that's no longer possible, I'll look for a new position, because I'm not prepared just to mark time.

**Poor answer to question 72**   Of course, I can. My friends say I'm very flexible.

**Good answer to question 72**   I've often adapted to new situations at work. There are always changes which need to be dealt with. My first position after completing my training was a case in point, as I moved from a small timber processing plant to a big building materials supplier. Over the years, I've had to learn new work processes and adapt to new roles.

73. Which events during your training / university course were decisive for you?

Your answer: _____

_____

_____

_____

_____

_____

_____

_____

_____

_____

74. Can you give us two examples of your ability to learn new things?

Your answer: _____

_____

_____

_____

_____

_____

_____

_____

_____

_____

_____

**Poor answer to question 73**   I failed an important exam. I felt like dropping out of my course at that time. It took a while before I applied to resit the exam.

**Good answer to question 73**   One key event was the customer focus I experienced during my work placement. I learned that I have an aptitude for working with customers, so I chose to go into sales and marketing after completing my course.

**Poor answer to question 74**   Yes. During my training I never had any problems revising for exams. And in my work I've never made the same mistake twice.

**Good answer to question 74**   Yes. Just last week I learned the latest version of the software that we use in our tax consultancy to draw up the year-end accounts. I also read up on special issues in taxation law, so that our clients can always turn to me for advice.

75. What new trends will we have to prepare for in our industry in the near future?

Your answer: _____

_____

_____

_____

_____

_____

_____

_____

_____

_____


76. What is different in your field of work now, compared to five years ago?

Your answer: _____

_____

_____

_____

_____

_____

_____

_____

_____

_____

**Poor answer to question 75**   Trends come and go. The main thing, in my opinion, is to offer good products.

**Good answer to question 75**   The consolidation process in the retail trade will continue to increase massively. That means that we're going to face ever more powerful competitors. On top of that, there are foreign retailers who are going to compete in our market. Of course, that presents an opportunity to compete in foreign markets ourselves.

**Poor answer to question 76**   Actually, not that much has changed. But globalisation is more of a curse than a blessing.

**Good answer to question 76**   Globalisation of the workplace has increased. I work a lot more with international business partners, these days. A large part of our correspondence in the company is in English.

77.  How do you react to criticism?

Your answer: _____

_____

_____

_____

_____

_____

_____

_____

_____

_____


78.  How do you go about criticising a superior, if he or she has
     made a clear mistake?

Your answer: _____

_____

_____

_____

_____

_____

_____

_____

_____

_____

**Poor answer to question 77**   Criticism is a fact of life, you have to get used to it.

**Good answer to question 77**   Criticism can be helpful if it's expressed in an objective way. I'm always open to hints about how to improve my work. If I think the criticism is unjustified, I will try to talk it through. That's usually enough to resolve any differences.


**Poor answer to question 78**   It's good for bosses to realise that they aren't perfect either. I would point out the mistake in front of the rest of the department.

**Good answer to question 78**   If I notice that something is going wrong, I talk to my manager in private. Of course, you have to pick the right moment. If my manager is stressed, that isn't a good time. I've noticed that managers appreciate it if employees point out mistakes in a calm and objective way.

79. What's your attitude towards failure management?

Your answer: _____

_____

_____

_____

_____

_____

_____

_____

_____

_____


80. What will you do if you don't get this position?

Your answer: _____

_____

_____

_____

_____

_____

_____

_____

_____

_____

**Poor answer to question 79**  It's just another management technique which has nothing to do with the reality of the workplace.

**Good answer to question 79**  It's important that everyone in the company works to eliminate errors as far as possible. Errors have a tendency to grow in scale the longer you leave them uncorrected. I also find it very unproductive to start blaming each other. It's better to have a zero-failure mindset.

**Poor answer to question 80**  Then I'll go elsewhere, there are other good jobs around.

**Good answer to question 80**  That would be a real pity, because I'd like to bring my professional experience to this position. As a personnel assistant I can offer experience in calculating expenses and commissions for the sales force in the field. I also have experience in maintaining personnel files, vacation and human resource planning and compiling statistical assessments. If I don't get this position, I'll apply for similar positions with other companies.

# 9

# How do you motivate yourself for work duties?

Ihr neuer Arbeitgeber möchte Ihre innere Einstellung zur täglichen Arbeit und Ihre Arbeitsmotivation kennen lernen. Die Personalverantwortlichen versuchen daher, mit verschiedenen Fragen herauszubekommen, ob Sie nur auf äußeren Druck reagieren oder sich durch Ihre innere Überzeugung leiten lassen.

### Hintergrund

Mitarbeiterinnen und Mitarbeiter, die sich mit ihren beruflichen Aufgaben identifizieren können, sind bei den Firmen gefragt. Denn motivierte Kandidaten zeichnen sich dadurch aus, dass sie sich selbst berufliche Ziele stecken, auf deren Erreichung hinarbeiten und besser mit Rückschlägen umgehen können als unmotivierte Kollegen. Zusätzlich geben diese gefragten Mitarbeiter ihrem beruflichen Umfeld positive Impulse: Andere Kollegen lassen sich von der Motivation anstecken, Arbeitsabläufe werden optimiert, und gemeinsam erreichte Ziele schweißen das Team zusammen.

### Typische Fehler

Viele Bewerber bezeichnen sich als motiviert, ohne dies näher begründen zu können. Es ist problematisch, wenn Bewerber Leerfloskeln und Schlagwörter herunterbeten, ohne sie mit Inhalt zu füllen und konkrete Beispiele zu nennen. Entsteht bei Personalverantwortlichen oder künftigen Fachvorgesetzten dann der Eindruck, dass die eigentliche Motivation nur darin besteht, am Monatsende ein Gehalt von der Firma überwiesen zu bekommen, ist der Kandidat im »Motivationstest« durchgefallen.

### Negativbeispiel

Wird eine Bewerberin im Vorstellungsgespräch mit der Frage *How do you motivate yourself in your daily work?* konfrontiert, wäre die folgende Antwort ungeeignet: *I'm always highly motivated in my work. It's important to me to always do my best. You can't achieve anything without motivation.*

### Kommentar zum Negativbeispiel

In dieser Beispielantwort reiht sich eine Nullaussage an die nächste. Die Personalverantwortlichen werden der Bewerberin nicht abnehmen, dass sie sich überhaupt jemals näher mit ihrer Arbeitsmotivation auseinandergesetzt hat. Verräterisch ist der letzte Satz: *You can't achieve anything without motivation.* Es entsteht der Eindruck, dass die Bewerberin dem Personalverantwortlichen nach dem Mund reden will. So kann sie aber nicht überzeugen, es fehlen glaubwürdige Beispiele, aus denen man die Motivation der Bewerberin heraushören und nachvollziehen könnte.

### Antwort-Strategie

Machen Sie in Ihrer Antwort deutlich, dass Sie schon immer über eine hohe Eigenmotivation verfügt haben. Begründen Sie kurz, warum Sie sich für Ihre Ausbildung beziehungsweise Ihr Studium entschieden haben. Dann sollten Sie anhand passender und überzeugender Beispiele erläutern, was Sie bei der Erledigung Ihrer beruflichen Aufgaben antreibt, woraus Sie Kraft schöpfen und dass Sie sich auch von Rückschlägen nicht unterkriegen lassen. Sie werden bei den Personalverantwortlichen zusätzlich punkten können, wenn Sie zudem klarmachen, dass Sie noch lange nicht zum Stillstand gekommen sind und sich beruflich – natürlich im Rahmen der neuen Stelle – immer weiterentwickeln möchten.

### Positivbeispiel

Die Frage *How do you motivate yourself in your daily work?* lässt sich dementsprechend besser so beantworten: *I motivate myself through good results. If I've completed a task successfully, that motivates me to tackle new tasks. For example, in my present company I got my daily routine running smoothly, then I took on a special assignment. The special assignment consisted of preparing sales figures and making them usable for marketing activities. I learned to use Excel in depth and presented my results as a PowerPoint presentation. Of course, I was doing my normal duties as a marketing assistant at the same time.*

81. What prompted your choice of training/university course?

Your answer: _____

_____

_____

_____

_____

_____

_____

_____

_____

_____

82. What motivates you in your daily work?

Your answer: _____

_____

_____

_____

_____

_____

_____

_____

_____

_____

**Poor answer to question 81**   I wasn't sure what I wanted to do. School doesn't really help you to make those kinds of decisions about your future career. So my choice was a bit random.

**Good answer to question 81**   At school I always had a strong interest in technical subjects / creative subjects / languages / science. I used my work placements to get a taste of different careers that might interest me and get my first real-world experience. I made my final decision after finding out about the career possibilities that training / a degree in ... would open up to me.

**Poor answer to question 82**   I tell myself that I have to pay the rent one way or another.

**Good answer to question 82**   I find it motivating to see things progressing. I like to set myself goals in my work. So I worked together with the customer service team to respond better to customers' wishes. It was a difficult task, but the positive feedback from customers encouraged me.

83. How do you deal with setbacks?

Your answer: _____

_____

_____

_____

_____

_____

_____

_____

_____

_____

84. What's really important to you?

Your answer: _____

_____

_____

_____

_____

_____

_____

_____

_____

_____

**Poor answer to question 83**   Sometimes, things don't go according to plan. You just have to get on with it. There are factors you can't control.

**Good answer to question 83**   Things don't always run smoothly. Setbacks are a sign that you have to approach something differently in the future. In our field sales team there was a problem with winning new customers. I helped to make sure that we prepared for appointments with customers by calling them on the phone and by sending them product information. After that we were able to grow our customer base considerably.

**Poor answer to question 84**   My health, my family and a reliable income.

**Good answer to question 84**   My family/friends and the opportunity to use my knowledge and experience in my work. I've always made a conscious effort to have a good grasp of my profession. That's why I undertook advanced training as a theatre nurse.

85. Why did you apply for this position?

Your answer: _____

_____

_____

_____

_____

_____

_____

_____

_____

_____


86. If you could start again, what would you do the same, and what would you do differently?

Your answer: _____

_____

_____

_____

_____

_____

_____

_____

_____

_____

**Poor answer to question 85**   My previous employer went out of business, so I had to look for another job.

**Good answer to question 85**   When I read your advertisement I was excited, because I identified with the job description. I have several years' experience in organising the back office. In my previous job I conducted a lot of correspondence in English. And I've often coached my colleagues in new software.

**Poor answer to question 86**   Well, I wouldn't listen to the people at the job centre. Nobody wants technicians these days. I would do a course in the commercial sector, then my future would be a lot more secure.

**Good answer to question 86**   By and large, I'm very happy with my situation. Perhaps I would do managerial training sooner, if I had my time again. On the other hand, I wouldn't want to miss out on my shop floor experience as a mechatronics engineer. It taught me a lot about fault diagnostics – knowledge I still use today in leading production teams.

87.  What personal and professional goals do you still want to fulfil?

Your answer: _____

_____

_____

_____

_____

_____

_____

_____

88.  What additional training would you like to do?

Your answer: _____

_____

_____

_____

_____

_____

_____

_____

**Poor answer to question 87**  My main goal is to get this job. In my personal life there are a lot things that could be improved, but I don't think that's relevant here.

**Good answer to question 87**  I have some professional goals I'd like to fulfil. For example, I can imagine extending my purchasing role into international purchasing. I have some experience of supplier integration, so a temporary placement abroad with a supplier would interest me. As far as my personal life is concerned, I'm happy with things as they are.

**Poor answer to question 88**  I'm not sure. I think that the importance of training courses is a bit exaggerated. I wouldn't mind learning a bit of Italian, I suppose. It would be useful on holiday.

**Good answer to question 88**  There are always new things to learn. In my spare time I've done courses in rhetoric, for example. It would be interesting to do a course in dealing with complaints. I'm also interested in learning more about sales promotion techniques to broaden my sales knowledge.

89. What are you proud of?

Your answer: _____

_____

_____

_____

_____

_____

_____

_____

_____

_____


90. What particularly appealed to you in the job advertisement?

Your answer: _____

_____

_____

_____

_____

_____

_____

_____

_____

_____

**Poor answer to question 89**   I'm proud of my son, he's doing really well at school.

**Good answer to question 89**   I'm proud of preventing operator errors in our workshop machinery. Because of the protective measures I suggested, incorrect operation is almost impossible now. I was also really pleased to be appointed to an interdepartmental quality management group.


**Poor answer to question 90**   I liked that the job is close to my home. I wouldn't want to relocate. The property market is so bad at the moment, I'd have to sell my house at a loss.

**Good answer to question 90**   It particularly appealed to me that you are looking for a marketing executive to work on the relaunch of your corporate identity and corporate design. I've already been involved extensively in those areas. For example, I helped to draw up a mission statement for my last employer and was involved in developing an umbrella campaign. I also have a good grasp of the other routine duties you mention in the advert.

91. Where do you want to be in five years' time?

Your answer: _____

_____

_____

_____

_____

_____

_____

_____

_____

_____


92. How long will you need to find your feet in this job?

Your answer: _____

_____

_____

_____

_____

_____

_____

_____

_____

_____

**Poor answer to question 91**   I haven't thought about that. I'll just be glad if I get the job.

**Good answer to question 91**   In five years' time I'd like to have a senior position in the central marketing office. I could also imagine having a managerial role in corporate communications. In my present job as a marketing assistant I've often taken on special assignments, so I can easily imagine taking on a more complex role.

**Poor answer to question 92**   It's certainly a big change for me. It isn't so easy to learn to get along with new colleagues. But I'm sure you'll find me a good worker, once I settle in.

**Good answer to question 92**   Some duties I'll be able to do straight away. That includes costing, account monitoring and soliciting tenders. When it comes to negotiating conditions and selecting suppliers, I would, of course, need some coaching from a colleague at first. I'm sure I can carry out all the duties you're looking for after a short orientation phase.

93. What would you do on your first day in the new job?

Your answer: _____

_____

_____

_____

_____

_____

_____

_____

_____

_____


94. Could you imagine accepting another position with us?

Your answer: _____

_____

_____

_____

_____

_____

_____

_____

_____

_____

**Poor answer to question 93**   First of all, I would find my way around the building. Then, I would just take things slowly step by step.

**Good answer to question 93**   I would go to my desk and introduce myself to my new colleagues. Then I would enquire  about the normal routine and ask my manager whether she has any specific tasks for me yet.


**Poor answer to question 94**   Well, if I don't stand a chance of getting the advertised position, I could do something else for you.

**Good answer to question 94**   The advertised position of human resources manager plays to my strengths most. However, I would consider a post as a recruitment manager, as I've worked in that area as well.

95. What conditions do you need to produce your best work?

Your answer: _____

_____

_____

_____

_____

_____

_____

_____

_____

_____

96. What's your attitude towards overtime?

Your answer: _____

_____

_____

_____

_____

_____

_____

_____

_____

_____

**Poor answer to question 95**   An understanding boss and nice colleagues are important to me.

**Good answer to question 95**   It's important to me that we're all pulling in the same direction and that I'm fully involved in work processes. It's also important, of course, to get the resources and information I need to do my job.

**Poor answer to question 96**   It's about time the government did something about it. Some people don't have a job, and others have more overtime than they can cope with. That isn't a healthy situation.

**Good answer to question 96**   I know that it's necessary to work longer hours sometimes. I'm prepared to do overtime when it's needed to get the job done. On one important commissioning project I worked around the clock with my team. It shouldn't be the norm, but sometimes you're faced with challenges like that.

# 10

# Do you have a realistic self-image?

Die Erfahrung zeigt, dass die Bewerberinnen und Bewerber, die über eine realistische Einschätzung ihrer eigenen Person verfügen, mit den Anforderungen am neuen Arbeitsplatz besser klarkommen als diejenigen, die unbedarft in ein neues Arbeitsverhältnis hineinstolpern. Daher werden Ihnen von den Personalverantwortlichen Fragen zu Ihrem Selbstbild gestellt und Ihre Antworten anschließend mit Kontrollfragen überprüft.

## Hintergrund

Bei den Fragen nach Ihrem Selbstbild geht es sowohl um die Einschätzung Ihrer individuellen beruflichen Stärken und Schwächen als auch darum, zu erfahren, welches Bild Sie von sich im Umgang mit anderen Menschen haben. Im Vordergrund steht also der Abgleich von Selbst- und Fremdbild. Um den Wahrheitsgehalt Ihrer Antworten zu überprüfen, kann es passieren, dass Sie mit möglichen Brüchen im Lebenslauf oder kritischen Formulierungen aus Arbeitszeugnissen konfrontiert werden.

## Typische Fehler

Personalverantwortliche würden Ihnen nicht glauben, wenn Sie behaupteten, dass Sie noch nie an Ihre eigenen Grenzen gestoßen seien oder niemals kleinere Reibereien mit Kollegen oder Vorgesetzten gehabt hätten. Gerade Bewerber, die in der letzten Stelle Schwierigkeiten mit Kollegen oder Vorgesetzten hatten, sollten aber bei Fragen zu diesem Thema darauf achten, sich nicht von ihren Emotionen überwältigen zu lassen. Sie sollten solche Fragen nicht dazu benutzen, um einmal richtig auszupacken und Dampf abzulassen. Zu viel Ehrlichkeit ist hier kontraproduktiv, denn damit können Sie bei Personalverantwortlichen kein Verständnis hervorrufen. Im Gegenteil, Sie werden selbst als Teil der geschilderten Probleme gesehen und gelten dann als schwieriger Mitarbeiter.

### Negativbeispiel

Bewerber sollten bei der Frage *How do you handle difficult colleagues?* nicht der Versuchung erliegen, endlich einmal abzurechnen. So darf die Antwort auf keinen Fall lauten: *Unfortunately, some of my colleagues are difficult to work with. It's bad enough if your managers don't support you, but if your colleagues are obstructive as well, it's impossible to work effectively.*

### Kommentar zum Negativbeispiel

Vielleicht hat der Bewerber tatsächlich schon öfter schwierige Situationen mit Kollegen und Vorgesetzten erlebt. Das Vorstellungsgespräch ist aber der falsche Platz, um auf diese Krisen der Vergangenheit ausführlich einzugehen. Ein Personalverantwortlicher wird aus dieser Antwort nur heraushören, dass der Bewerber öfter mit Vorgesetzten aneinandergerät und Probleme mit Kollegen eskalieren lässt. Zudem neigt der Kandidat zu der negativen Eigenschaft der Schuldverschiebung: Nie ist er selbst schuld, immer sind es die anderen gewesen.

## Antwort-Strategie

Zeichnen Sie ein realistisches Bild von sich. Schwierige Fachaufgaben und persönliche Unstimmigkeiten gehören zum Berufsalltag mit dazu. Anstatt zu behaupten, noch nie an die eigenen Grenzen gestoßen zu sein oder kleinere Streitigkeiten mit Kollegen oder Vorgesetzten gehabt zu haben, sollten Sie lieber Ihre Fähigkeit herausstellen, in kritischen Situationen Lösungen entwickeln zu können. Stellen Sie sich als konstruktiven Menschen dar, der weiß, dass die tägliche Arbeitswelt nicht immer rosarot gefärbt ist. Überlegen Sie sich vor dem Gespräch genau, welche (ausgewählten) Schwierigkeiten Sie erwähnen, und vor allem, wie Sie sie aus der Welt geschafft haben. Schildern Sie nur typische, kleinere Unstimmigkeiten, die jeder schon einmal erlebt hat. Außerdem sollten Sie sich vorab auch überlegen, wie Sie eventuelle Brüche im Lebenslauf plausibel erklären können.

## Positivbeispiel

Zeigen Sie, dass Sie über ein realistisches Selbstbild verfügen, indem Sie die Frage *How do you handle difficult colleagues?* beispielsweise so beantworten: *Sometimes difficult situations with colleagues occur and need to be sorted out. Usually, you can do that if you try to see things from the colleague's perspective. I've always been able to do that. For example, there was a disagreement in our team about how tasks were shared out. I knew which tasks my colleagues preferred, so we eventually found a compromise that everybody could accept.*

## Kommentar zum Positivbeispiel

Die Antwort ist sehr geschickt, da Personalverantwortliche dem Bewerber gelebte Teamfähigkeit zurechnen werden. Dieser Bewerber weiß, dass im Berufsalltag Kompromisse im Umgang miteinander erforderlich sind. Es ist ihm klar, dass Menschen unterschiedliche Vorstellungen haben, die immer wieder einmal miteinander abgeklärt werden müssen. Mit dieser gelungenen Antwort ist der Bewerber aber nicht mehr Teil des Problems, sondern der Lösung. Er liefert einen Beleg dafür, dass er aktiv auf Kollegen zugehen und sie auf ein gemeinsames Vorgehen einschwören kann.

97. What do you do when you're at a loss for a solution?

Your answer: _____

_____

_____

_____

_____

_____

_____

_____

_____

_____

98. Where do you see shortcomings in yourself that you need to
    work on?

Your answer: _____

_____

_____

_____

_____

_____

_____

_____

_____

_____

**Poor answer to question 97**   That never happens, I always think of something. If necessary, my colleagues have to help me out.

**Good answer to question 97**   I find out about different ways of approaching the task. I ask colleagues for advice. Sometimes, it's a good idea to get information from other departments as well. If I couldn't get any information, I would turn to my line manager for help.

**Poor answer to question 98**   Nobody's perfect. I would like to be more open with people. Sometimes, I think I'm too much of a pessimist. And I could do with losing a few kilos.

**Good answer to question 98**   I don't think I have any major shortcomings. I'd be interested in doing a Spanish language course, and I'd like to do more training in rhetoric to become better at making speeches off the cuff.

99. What are your strengths and weaknesses?

Your answer: _____

_____

_____

_____

_____

_____

_____

_____

_____

_____

100. How will you approach your new colleagues?

Your answer: _____

_____

_____

_____

_____

_____

_____

_____

_____

_____

_____

**Poor answer to question 99**  I have a good sense of what is achievable. My particular strengths are positive thinking, optimism without naivety and commitment. My weaknesses include the fact that I can be direct and stubborn. I'm always honest, but sometimes I'm not diplomatic enough.

**Good answer to question 99**  My strengths include teamwork. I have a good understanding of the processes involved in product management and know how I can best use the talents of the people involved. When there's a heavy workload, I can motivate others by making sure they understand how important their contribution is to the team's results. In addition, my good head for figures has always helped me to draw the right conclusions from market research. My weakness is that I'm a bit too direct, sometimes. I need to learn that departmental diplomacy is important to get a project started.

**Poor answer to question 100**  I hope that my new colleagues will like me and won't be difficult.

**Good answer to question 100**  I'll try to establish a personal connection with each of my colleagues. That leads to better teamwork. Everyone has their favourite subjects that they like to talk about. I'll find out how things work and then help to get the job done.

101. What annoys you most about other people?

Your answer: _____

_____

_____

_____

_____

_____

_____

_____

_____

_____

102. Tell us about a line manager / training supervisor you had difficulties with.

Your answer: _____

_____

_____

_____

_____

_____

_____

_____

_____

_____

**Poor answer to question 101**   Some people are real tyrants and stifle everyone's initiative. My last boss was like that. He was so arrogant that he never listened to anyone else's opinion.

**Good answer to question 101**   Everyone has their peculiarities, and you have to get used to them. But I don't like it when people withhold information or give out false information. You can't really work with that kind of person.

**Poor answer to question 102**   I had a boss who was completely unreliable. He'd make big promises, then nothing would come of it. My training supervisor had his problems, too. Mostly with alcohol.

**Good answer to question 102**   I've always got on well with my line managers and with my training supervisor. With some of them it took more effort, but I always managed. To take my team leader in assembly, as an example, if his football team had lost at the weekend, you couldn't speak to him on Monday morning. You just had to work with that.

103. What do you expect from your new line manager?

Your answer: _____

_____

_____

_____

_____

_____

_____

_____

_____

_____


104. What do you expect from your new team?

Your answer: _____

_____

_____

_____

_____

_____

_____

_____

_____

_____

_____

_____

**Poor answer to question 103** Mostly, I expect her to back me up.

**Good answer to question 103** I want her to involve me in the work of the department. In a new job it's important to find out who is responsible for what. That's where I'd like support from my line manager.

**Poor answer to question 104** I've had mixed experiences in the past. I hope that things go better this time and it's a team I can work with.

**Good answer to question 104** I expect them to be willing to co-operate with me. I've always found it worthwhile to consult actively with my new colleagues. That way you get a real team spirit and the work runs a lot more smoothly.

105. Which colleagues do you most enjoy working with?

Your answer: _____

_____

_____

_____

_____

_____

_____

_____

_____

_____


106. How would you describe your work style?

Your answer: _____

_____

_____

_____

_____

_____

_____

_____

_____

_____

_____

**Poor answer to question 105**  I generally prefer working with younger colleagues. It's often difficult to make a connection with older colleagues, they're so set in their ways.

**Good answer to question 105**  I most enjoy working with colleagues who want to achieve something. The best colleagues to work with are the ones who have the company's interests at heart. I've always been able to get along well with my colleagues so far.

**Poor answer to question 106**  My motto is: get the job done as fast as required, but don't attempt the impossible.

**Good answer to question 106**  Fast, effective and proactive. I start by categorising my tasks according to urgency. I discuss the more complex tasks with the other people involved, and I prepare thoroughly for meetings. I think it's important to keep my colleagues informed and deliver my results on time.

107. When did you last lose your temper and why?

Your answer: _____

_____

_____

_____

_____

_____

_____

_____

_____

_____

108. What really makes you angry about other people?

Your answer: _____

_____

_____

_____

_____

_____

_____

_____

_____

_____

**Poor answer to question 107**   You wouldn't believe the questions I get asked in interviews. Recently, an interviewer asked me whether I was too old for the job. What a cheek!

**Good answer to question 107**   I rarely lose my temper, but I do get annoyed sometimes. The last time I was angry it was because of a trainee who just wanted to do his time with us and couldn't be bothered to do any work.


**Poor answer to question 108**   Stupidity. Unfortunately there's a lot of it around.

**Good answer to question 108**   Malicious or destructive behaviour. Deception or misinformation on the part of colleagues would make me angry.

109. What do your colleagues value about you?

Your answer: _____

_____

_____

_____

_____

_____

_____

_____

_____

_____


110. How do you deal with difficult people?

Your answer: _____

_____

_____

_____

_____

_____

_____

_____

_____

_____

**Poor answer to question 109**   No idea. You'd have to ask my colleagues.

**Good answer to question 109**   My colleagues would say that I'm a good listener and I'm always ready to help with queries or problems. They would also say that I'm reliable and I don't leave anyone out in the cold. On the contrary, I've always looked after my colleagues, whether it's been a question of cover for leave or taking on extra duties.

**Poor answer to question 110**   I try to avoid them.

**Good answer to question 110**   You have to get along with everyone, even difficult people. When I'm dealing with customers, I make a point of handling the difficult ones well. Often they only appear difficult at first: there's usually some way of getting through to them. It's better to adapt yourself to customers than to try to force them to do things your way. One customer may be tempted by a good price, another by technical features and a third by our strong brand.

111. What things do you need in order to work productively?

Your answer: _____

_____

_____

_____

_____

_____

_____

_____

_____

_____

112. How would your present boss describe you?

Your answer: _____

_____

_____

_____

_____

_____

_____

_____

_____

**Poor answer to question 111**   All I need to do my job is some peace and quiet.

**Good answer to question 111**   I need to be well integrated into the company to work productively. All the team members need to be pulling in the same direction. And there needs to be the opportunity to make my own contribution.

**Poor answer to question 112**   I don't really know. After all, I want to leave the company. He may say bad things about me just because I've left.

**Good answer to question 112**   He would say that I'm reliable. He would also mention that I work well with my colleagues. Perhaps he would also talk about the success of our new sales promotion, to which I was a significant contributor.

# 11

# How do you deal with conflict?

Nicht wenige Personalverantwortliche sind der Überzeugung, dass Menschen erst dann ihr wahres Gesicht zeigen, wenn der Wind etwas rauer wird, wenn also zwischenmenschliche Konflikte auftreten. Daher werden in Vorstellungsgesprächen neuerdings auch spezielle Fragen zum Konfliktverhalten der Bewerber gestellt. Personalverantwortliche wollen herausfinden, wie die Kandidaten mit Meinungsverschiedenheiten, Belastungen, Enttäuschungen oder sonstigen Konfliktsituationen umgehen.

## Hintergrund

Bei der Arbeit lässt sich der Faktor Mensch nie völlig ausblenden. Ganz im Gegenteil wird es im Zeitalter abteilungsübergreifender Projektarbeit immer wichtiger, sich in kürzester Zeit auf unterschiedlichste (Fach-)Spezialisten einzustellen – und die haben nun einmal oft ihre persönlichen Eigenarten. Hinzu kommt, dass knappe Terminvorgaben und enge Zeitressourcen die Belastung jedes Einzelnen noch weiter erhöhen. Und auch der Umgang mit dem Chef ist nicht immer konfliktfrei, sodass ein konstruktives Konfliktverhalten in der heutigen Arbeitswelt immer wichtiger wird.

## Typische Fehler

Wer von sich behauptet, konfliktstark zu sein, gleichzeitig aber an früheren Kollegen kein gutes Haar lässt oder an ehemaligen Vorgesetzten herumnörgelt, fällt in diesem Fragenkomplex durch. Auch derjenige, der keine Beispiele dafür liefern kann, dass er sich mit sachlicher Kritik vonseiten anderer gründlich auseinandersetzt, lässt Zweifel an seinem Konfliktverhalten aufkommen. Besonders gefürchtet sind zudem diejenigen Bewerberinnen und Bewerber, die die Gründe für berufliche oder private Fehlentwicklungen stets zuerst bei anderen und zuletzt bei sich selbst suchen.

## Negativbeispiel

Ungünstig wäre es, wenn ein Bewerber auf die Frage *Are you able to handle conflict?* folgendermaßen antwortet: *Of course. If anyone has a problem with me, they can tell me so face to face. I have years of experience in my field, nobody can make a fool of me. I've never been afraid to speak my mind, and I never will.*

## Kommentar zum Negativbeispiel

Hier verwechselt der Bewerber Konfliktfähigkeit mit Durchsetzungsvermögen. Nicht jeder Konflikt lässt sich aber so auflösen, dass man auf seiner Position beharrt und dem Gegenüber die Stirn bietet. Der Rückzug auf die fachliche Autorität verdeutlicht dem Personalverantwortlichen, dass er hier einen Bewerber vor sich hat, der zwar ein fachlich guter Spezialist sein mag, der aber Schwierigkeiten haben wird, sich erfolgreich in ein Team zu integrieren.

## Antwort-Strategie

Fragen zum Konfliktverhalten werden Sie dann mit Bravour meistern, wenn Sie typische Konfliktsituationen aus Ihrem Berufsfeld nennen können und gleichzeitig erläutern, wie Sie sie aufgelöst haben. Zeigen Sie, dass Sie vor Schwierigkeiten nicht weglaufen, sondern bereit sind, sich unangenehmen Situationen zu stellen. Betonen Sie Ihre Fähigkeit, nach Kontroversen wieder auf andere zugehen zu können, um gemeinsam konstruktive Lösungen zu entwickeln. Heben Sie hervor, dass Sie im Allgemeinen gut mit Kollegen und Vorgesetzten zurechtkommen, dass Sie aber dann, wenn es einmal zum Konflikt kommt, geeignete Lösungsmaßnahmen einsetzen.

## Positivbeispiel

Eine gelungene Antwort auf die Frage *Are you able to handle conflict?* könnte demnach so lauten: *There are always differences of opinion at work. The important thing is to use them productively. In the sales department there was a decision to introduce call flow scripts. We in the field sales team thought that the parameters were too rigid. After discussions with the sales managers and the training manager, we reached a compromise: experienced colleagues were allowed more flexibility, and beginners were given the scripts to help them find their feet.*

## Kommentar zum Positivbeispiel

Der Bewerber zeigt in diesem Beispiel, dass er fähig ist, schwierige Situationen konstruktiv zu lösen: Er schildert einen typischen beruflichen Konflikt aus seinem Arbeitsfeld. Seine Darstellung der Auflösung dieses Konfliktes ist gelungen, denn starren Haltungen, die zu einer Verhärtung der Fronten geführt hätten, gibt der Bewerber keine

Chance. Alle Beteiligten können bei dieser Lösung ihr Gesicht wahren: Die Verantwortlichen für die Einführung des neuen Gesprächsleitfadens können ihre Absichten dort verwirklichen, wo es wirklich sinnvoll ist – nämlich bei Berufseinsteigern, und die berufserfahrenen Kollegen erhalten weiterhin Freiräume, die sie sinnvoll ausfüllen können, und müssen sich nicht an starre Vorgaben halten.

113. How do you deal with disappointments in your work?

Your answer: _____

_____

_____

_____

_____

_____

_____

_____

_____

_____

114. Do you think you have been given enough opportunities in
     your present job?

Your answer: _____

_____

_____

_____

_____

_____

_____

_____

_____

_____

**Poor answer to question 113**   There's nothing you can do. Disappointments are a fact of life. Unfortunately, the workers get blamed for situations that are actually someone else's fault.

**Good answer to question 113**   I haven't experienced any real disappointments in my work. Of course, things don't always turn out perfectly. I thought it was a pity that my line manager didn't release me for the umbrella marketing project group. But I didn't give up, and now I'm involved in an interdepartmental sales promotion project.

**Poor answer to question 114**   I could have achieved a lot more if my boss had been more supportive. That's why I want to leave the company.

**Good answer to question 114**   It's up to people to make their own opportunities at work. I've taken the initiative with special assignments and projects. Of course, it's important to my line manager that the work in the department takes priority, but I was able to reassure her that I would continue to do good work for her and also raise the department's profile in the company.

115. Do you think that your present boss has recognised your professional potential?

Your answer: _____

_____

_____

_____

_____

_____

_____

_____

_____

_____


116. What things have bothered you in your present job? And what have you done to deal with them?

Your answer: _____

_____

_____

_____

_____

_____

_____

_____

_____

_____

**Poor answer to question 115**   I sometimes have the impression that there's too much going on in our department. In all events, my boss seems to be out of his depth, he doesn't pay much attention to his employees.

**Good answer to question 115**   My manager knows that he can rely on me. I've often taken on special assignments for him. I can tell from that, that he recognises my potential.

**Poor answer to question 116**   I don't think about it. Things aren't great, but that's the way work is. There's nothing you can do about it.

**Good answer to question 116**   When I started in my last job, there wasn't much dialogue between us in the service department and the people in development. I spoke to my manager about it. He encouraged us to establish lines of communication. We managed to set up a meeting every two weeks to swap experiences.

117. What don't you like about your present line manager?

Your answer: _____

_____

_____

_____

_____

_____

_____

_____

_____

_____

118. How do your colleagues know when your patience is
exhausted?

Your answer: _____

_____

_____

_____

_____

_____

_____

_____

_____

_____

**Poor answer to question 117**   There are a few things. He's quite impatient, even hot-headed. Even though the personnel department sent him on a management course.

**Good answer to question 117**   I get on well with my line manager. He has his idiosyncrasies, like everyone. But you can make allowances for that. He tends to be quite demanding, but that benefits the department in the long run.

**Poor answer to question 118**   I don't know really. When it's all too much, I get very stubborn. That surprises my colleagues.

**Good answer to question 118**   I think that you should tell colleagues when you aren't happy with how things are going. Just waiting for them to realise that there's a problem isn't enough, and it is detrimental to the company.

119. What would the people in your present team criticise about you?

Your answer: _____

_____

_____

_____

_____

_____

_____

_____

_____

120. How do you deal with criticism?

Your answer: _____

_____

_____

_____

_____

_____

_____

_____

_____

**Poor answer to question 119**   Not a lot, I hope. But you never really know what your colleagues think of you.

**Good answer to question 119**   Perhaps that I don't like to discuss the same point ten times. I know that it's important to consult people, but I do like things to keep moving forward.

**Poor answer to question 120**   In an open-minded, honest way. That's what's expected.

**Good answer to question 120**   I listen to the criticism carefully. It can be helpful. It needs to be given in a constructive way, though. If I think the criticism isn't justified, I try to discuss the matter with the person in private. Most ill-feeling can be diffused in that way.

121. How do you react if you are criticised unfairly?

Your answer: _____

_____

_____

_____

_____

_____

_____

_____

_____

_____

122. How do you deal with uncomfortable situations?

Your answer: _____

_____

_____

_____

_____

_____

_____

_____

_____

_____

**Poor answer to question 121**  Then I stand up for myself and take the offensive.

**Good answer to question 121**  I try to dispel the criticism. Preferably in a private conversation in which I try to find out what the real problem is. Of course, I don't just accept unjustified criticism.

**Poor answer to question 122**  I try to avoid uncomfortable situations.

**Good answer to question 122**  You have to face up to uncomfortable situations. I remember how uncomfortable I felt when I had to deal with my first customer complaint. There are still some situations I find more difficult than others. My professional experience helps me to deal with them successfully.

123. Can you think of a situation in which you had a difference of opinion with a colleague and explain how you resolved it?

Your answer: _____

_____

_____

_____

_____

_____

_____

_____

_____

_____


124. Under what circumstances do you enjoy your daily work?

Your answer: _____

_____

_____

_____

_____

_____

_____

_____

_____

_____

**Poor answer to question 123**   Oh yes, lots of them. I always get back-up from my boss. I've always managed to keep my colleagues under control that way.

**Good answer to question 123**   When we introduced a customer relationship management system in our department, a lot of people had reservations about it, because reports had to be entered twice. My colleagues protested at this extra work and simply ignored the new system. I wasn't too happy with the duplicated work either, but I said that we should talk to the IT department about it and offered to act as the contact person. The developers managed to program a tool for us which avoided the unnecessary work.

**Poor answer to question 124**   When everything is running smoothly, my colleagues and boss are in a good mood and there isn't too much work.

**Good answer to question 124**   I enjoy my work when I can achieve something. It could be a specific goal we've set, or, more generally, a positive outcome for the company.

125. What makes you angry in your daily work?

Your answer: _____

_____

_____

_____

_____

_____

_____

_____

_____

_____

126. With which personality traits of your colleagues do you have real difficulty?

Your answer: _____

_____

_____

_____

_____

_____

_____

_____

_____

_____

**Poor answer to question 125**  Pointless directives from above. The managers often have no idea of the reality on the shop floor.

**Good answer to question 125**  When personal feelings become more important than the actual work. It annoys me when people prevent the team from getting results by their obstructive behaviour.

**Poor answer to question 126**  I think that some of my colleagues just want to annoy me. Some are deliberately obtuse, others act like they know it all. It's really quite difficult to keep your cool with colleagues like that.

**Good answer to question 126**  I get on well with my colleagues. Some of them have their idiosyncrasies, but you can make allowances for that. You have to know how to get along with your colleagues and when is a good time to bring up a difficult topic.

127. What would make you leave our company?

Your answer: _____

_____

_____

_____

_____

_____

_____

_____

_____

_____

_____


128. Who do you turn to for advice in potentially difficult situations?

Your answer: _____

_____

_____

_____

_____

_____

_____

_____

_____

_____

_____

**Poor answer to question 127**   If the money wasn't right.

**Good answer to question 127**   If there was no longer any need for my contribution or the company was in a precarious financial position, I suppose I'd have to look for a new job.

**Poor answer to question 128**   I can manage by myself just fine, I prefer to deal with situations my own way.

**Good answer to question 128**   It depends on the situation. If it's a technical issue, there are always specialists to help me, even if I have to go outside my department. If it's a personal difficulty, then I'm confident that I can sort it out for myself. Sometimes, though, it's useful to talk things over with colleagues or friends.

# 12

# How do you deal with stress questions and unlawful questions?

Bei Stressfragen ist die Firmenseite häufig nur in zweiter Linie an der eigentlichen Antwort des Bewerbers interessiert. An erster Stelle steht vielmehr die Art und Weise, *wie* der Bewerber antwortet. Das Gleiche gilt auch für den Umgang mit eigentlich nicht zulässigen Fragen, die von Personalverantwortlichen ebenfalls als Stressfragen eingesetzt werden können und in einigen speziellen Fällen auch gestellt werden dürfen – dazu gleich mehr.

## Hintergrund

Auch wenn Bewerberinnen und Bewerber häufig das gesamte Vorstellungsgespräch als permanenten Stress empfinden, ist nicht jede Frage auch gleich eine Stressfrage. Echte Stressfragen werden aus verschiedenen Gründen gestellt. Personalverantwortliche setzen sie beispielsweise ein, wenn die bisherigen Antworten der Bewerber nicht überzeugen konnten und jetzt durch gezieltes Nachfragen noch einmal überprüft werden sollen. So manche Stressfrage wird aber auch eingestreut, um zu sehen, wie der Bewerber auf ungewöhnliche Fragen reagiert oder mit zusätzlichem Druck umgeht.

## Unzulässige Fragen

Fragen zu Kinderwunsch, Schwangerschaft, Vorstrafen, Lohnpfändungen oder zu Konfessions-, Partei- oder Gewerkschaftszugehörigkeit sind grundsätzlich unzulässig, und Sie müssen dann in der Regel auch nicht wahrheitsgemäß antworten. Sie dürfen aber dann gestellt werden, wenn die Information für die zukünftige Arbeit unabdingbar ist. Beispielsweise ist die Frage nach einer bestehenden Schwangerschaft erlaubt, wenn mit fruchtschädigenden Substanzen im Labor gearbeitet werden soll. Wenn der Arbeitgeber ein so genannter »Tendenzbetrieb« ist – also ein kirchlicher Träger, eine Parteistiftung, ein Arbeitgeberverband oder ein Gewerkschaftsbund –, sind Fragen nach einer entsprechenden Mitgliedschaft zulässig.

## Typische Fehler

Stressfragen werden von Personalprofis gerne als »kleiner Soft-Skill-Test« genutzt. Die Reaktionen der Bewerber zeigen schnell, wie es um ihre angeblich vorhandenen Soft Skills wie beispielsweise *Belastbarkeit*, *Kommunikationsstärke* oder auch *Konfliktfähigkeit* in der Praxis bestellt ist. Wer deshalb auf Stressfragen patzig reagiert, nur noch trotzig schweigt oder kämpferisch betont, dass die Frage schon aus arbeitsrechtlichen Gründen unzulässig sei, stellt sich selbst ins Abseits. Denn wenn zwischen dem Bewerber und den an der Einstellungsentscheidung beteiligten Personen auf der Firmenseite erst einmal Kampfstimmung aufgekommen ist, ist der Kandidat aus dem Rennen.

## Negativbeispiel

Eine Bewerberin bewirbt sich um eine Stelle im Verkauf und wird vom Personalverantwortlichen mit folgender unzulässiger Frage konfrontiert: *Do you want to have children?* Ihre folgende Antwort ist dann ungünstig: *As you know, that question is unlawful, so I'm not going to answer it.*

## Kommentar zum Negativbeispiel

Die Bewerberin ist zwar im Recht, sollte sich aber dennoch versöhnlicher geben. So erweckt sie den Eindruck, dass sie zur Sturheit neigt, und das wiederum könnte kontraproduktiv im Umgang mit Kunden sein. Denn auch die werden ab und an einmal unsinnige Fragen stellen, auf die es gelassen zu reagieren gilt.

## Antwort-Strategie

Zeigen Sie mit Ihrem Antwortverhalten, dass Sie sich nicht so schnell aus der Ruhe bringen lassen. Reagieren Sie auf Provokationen, Suggestivfragen oder Unterstellungen nicht mit Kampfrhetorik. Unfaire Angriffe seitens der Personalprofis laufen ins Leere, wenn Sie Ihr diplomatisches Geschick einsetzen und geduldig und freundlich antworten. Zeigen Sie Ihren Gesprächspartnern noch einmal, dass Sie wissen, was Sie beruflich können und was Sie wollen. Sie sollten auch Ihre Körpersprache gezielt einsetzen, um Ihre Souveränität zu unterstreichen. Halten Sie bei Ihren Antworten Blickkontakt zu den Fragestellern und lehnen Sie sich im Stuhl immer wieder einmal zurück, um körperliche Verspannungen aufzulösen.

## Positivbeispiel

Souveräner könnte die Bewerberin auf die Frage *Do you want to have children?* folglich mit dieser Antwort reagieren: *That isn't an issue for me, right now. I want to work in your sales department and advise your customers. That's my top priority.*

## Kommentar zum Positivbeispiel

Diese Antwort ist viel geschickter als die schroffe Zurückweisung im Negativbeispiel. Kurz und knackig hakt die Bewerberin die Stressfrage ab und behält dabei gleichzeitig die Gesprächssituation im Blick – schließlich geht es darum, zu überprüfen, ob sie wirklich eine Arbeit im Verkauf annehmen möchte. Da die Bewerberin dies bejaht und ihre Freude an der Kundenberatung noch einmal betont, wird der Personalverantwortliche zufrieden sein.

129. Are you sure you're the right person for the job?

Your answer: _____

_____

_____

_____

_____

_____

_____

_____

_____

_____


130. You still haven't convinced me: why should we give you the job?

Your answer: _____

_____

_____

_____

_____

_____

_____

_____

_____

_____

**Poor answer to question 129**   Why else would I be here?

**Good answer to question 129**   I'm quite sure that I'll be able to handle my new responsibilities. In my last job I was responsible for achieving market objectives, such as winning new customers and controlling distribution channels. I've informed myself about your products, and I'm convinced that I'll have further market successes with you.

**Poor answer to question 130**   I'm sorry that you feel that way. I really can't say much more about myself. I've described what I did in my last job. I'm not the only person with experience in purchasing, after all.

**Good answer to question 130**   Because I have wide-ranging experience in international purchasing. Also, I have a very good knowledge of the supplier market. With my last employer I reduced production costs by better supplier integration. Also, I achieved massive savings in purchasing. I'd like to achieve similar successes in your company.

131. You didn't stay with your last employer long. How do we know you won't leave us after a short time, too?

Your answer: _____

_____

_____

_____

_____

_____

_____

_____

_____

_____

132. Just between us: why do you really want to leave your present employer?

Your answer: _____

_____

_____

_____

_____

_____

_____

_____

_____

_____

**Poor answer to question 131**   It wasn't my fault. My manager was difficult to get along with. He always rejected my new ideas, which meant that I had no scope for my creative potential.

**Good answer to question 131**   You're right: I only stayed in my last position for eight months. But before that I was in the same job for four years. I would have stayed longer with my last employer, too. But due to internal restructuring my job was at risk, and that made me decide to look for another employer.

**Poor answer to question 132**   I have to tell you, it's complete chaos in my company at the moment. The right hand doesn't know what the left hand is doing. It surprises me that things have gone so well for so long. Now we're getting a new manager on top of everything, so I think it's time for me to go.

**Good answer to question 132**   I have a lot of respect for my present employers. I've been able to further my professional development with them. But it's important to me to use my experience now in another setting and another company. I'd like to get off to a flying start with the five years' professional experience that I've accumulated.

133. Won't you be planning a family in the near future?

Your answer: _____

_____

_____

_____

_____

_____

_____

_____

_____


134. Won't you be out of your depth in this position?

Your answer: _____

_____

_____

_____

_____

_____

_____

_____

_____

_____

**Poor answer to question 133**   I don't think you need to worry about that. If I do become pregnant, I'll still manage somehow. Anyway, you aren't allowed to ask that question.

**Good answer to question 133**   I've discussed that with my partner, and we're agreed that our careers are the main priority. If I can continue my career with you, I'll be very happy.

**Poor answer to question 134**   Well, you have to be optimistic. I'm sure I'll manage.

**Good answer to question 134**   A lot of the duties you've described to me are ones I'm already familiar with from my previous jobs. So I know what to expect, and I'm looking forward to my new role.

135. Isn't the advertised job a bit of a step down for you?

Your answer: _____

_____

_____

_____

_____

_____

_____

_____

_____

_____

136. Between you and me: they asked you to resign, didn't they?

Your answer: _____

_____

_____

_____

_____

_____

_____

_____

_____

_____

**Poor answer to question 135**   At my age there aren't so many suitable jobs. Of course, I'm capable of more, but perhaps I'll be able to persuade you to let me take on a more responsible role in due course.

**Good answer to question 135**   I don't see it like that. In the new position I'll have interesting tasks to work on. I've had a variety of jobs in the course of my career. I'm sure I'll be happy in my new position and with my new role.

**Poor answer to question 136**   Well, I prefer to jump before I'm pushed.

**Good answer to question 136**   No, my company doesn't know anything about my plans yet. I took a day's leave for this interview. I could easily stay with my present employer. But this new job interests me, because it gives me the chance to take on more responsibility.

137. What things in your life are so important that they take precedence over your work?

Your answer: _____

_____

_____

_____

_____

_____

_____

_____

_____

_____

138. What has been your biggest mistake in life?

Your answer: _____

_____

_____

_____

_____

_____

_____

_____

_____

_____

**Poor answer to question 137**   My health.


**Good answer to question 137**   I enjoy my work and also make active use of my leisure time. So I can't really imagine anything that might not be compatible with my career.


**Poor answer to question 138**   I should have gone to university, but times were different back then. My parents wanted me to learn a trade first. I found myself stuck on this career path. Now it's too late to go back.

**Good answer to question 138**   I'm happy with my life as it is. I can't think of a really big mistake. Perhaps it would have been interesting for me to work abroad for a while. But that opportunity hasn't come up yet.

139. Couldn't you have achieved your career goals more quickly?

Your answer: _____

_____

_____

_____

_____

_____

_____

_____

_____

_____


140. Imagine you had 60 days holiday a year. What would you do with that time?

Your answer: _____

_____

_____

_____

_____

_____

_____

_____

_____

_____

**Poor answer to question 139**   If I had known before what I know now, I would have achieved more in life. You have to shout if you want to be noticed. If I had made wild promises like some of my colleagues, I might have got further in life.

**Good answer to question 139**   Not really, unless I'd had a lot of luck. I've worked steadily on my career, and I'm very happy with what I've achieved.

**Poor answer to question 140**   I would catch up with some things that need doing. My house could do with some renovations, and my wife has been nagging me about a conservatory for years.

**Good answer to question 140**   There are some things that need doing at home, and then I'd learn another foreign language, perhaps Spanish. I could do a language course in Spain or Mexico, then it would be a bit of a holiday as well.

141. A lot of employees are afraid to contradict their boss. What do you think of that attitude?

Your answer: _____

_____

_____

_____

_____

_____

_____

_____

_____

142. If you won a million euros in the lottery, what would you do?

Your answer: _____

_____

_____

_____

_____

_____

_____

_____

_____

_____

**Poor answer to question 141**   There are a lot of conformists and yes-men in the workplace. Not many people dare to contradict their boss. The reason, though, is that most managers don't take criticism well.

**Good answer to question 141**   Some people may feel that way, but I think it's better to have a good working relationship with my manager. In my department we work together well, including the manager.

**Poor answer to question 142**   I'd give up work.

**Good answer to question 142**   I don't think it would change my life very much at all. Of course, I'd be happy to have the money for a property in a good location. My children would also get something. But I would get on with my life as normal.

143. Why weren't you able to turn things around at your old firm?

Your answer: _____

_____

_____

_____

_____

_____

_____

_____

_____

144. Do you prefer to shower or take a bath?

Your answer: _____

_____

_____

_____

_____

_____

_____

_____

_____

**Poor answer to question 143**   Sometimes people just won't listen to good advice. Then they have to learn the hard way.

**Good answer to question 143**   I did everything in my power to suggest improvements and make changes. Our department continued to run well until the end. Everything else was out of my hands.

**Poor answer to question 144**   You want to know whether I'm wasteful or economical? Then I'll choose the shower.

**Good answer to question 144**   I don't have much time to spare in the mornings, so I take a shower.

# 13

# Are you able to dispel prejudice?

Es gibt Bewerbergruppen, die es in Vorstellungsgesprächen schwerer haben als andere. Das gilt beispielsweise für Bewerber, die häufiger den Arbeitgeber gewechselt haben, für jüngere, ältere oder arbeitslose Bewerber sowie für Bewerberinnen, die nach einer Erziehungszeit wieder ins Berufsleben zurückkehren möchten.

## Hintergrund

Wenn Personalverantwortliche davon ausgehen, dass Bewerber in der zu vergebenden Stelle mit bestimmten Vorurteilen fertig werden müssen, nehmen sie diese kritischen Situationen gerne vorweg. Dann werden die Bewerber bereits im Vorstellungsgespräch mit den Vorurteilen konfrontiert, mit denen sie auch am späteren Arbeitsplatz kämpfen werden müssen. Die Reaktionen der Bewerber auf die ihnen gegenüber geäußerten Vorurteile und ihre Fähigkeit, diese Voreingenommenheiten zu entkräften, sind für die Personalverantwortlichen dann ein Einstellungskriterium.

## Typische Fehler

Wer sich auf Grundsatzdiskussionen über Vorurteile einlässt, hat schon verloren. Die Klischees vom jungen Bewerber, der zu wenig Berufserfahrung hat, vom alten Bewerber, der nicht mehr bereit ist, sich auf Neues einzustellen, vom arbeitslosen Bewerber, der den Anschluss nicht mehr findet, oder auch vom Bewerber aus den neuen Bundesländern, der Schwierigkeiten mit seinen »Ostwurzeln« hat, sollten Sie auf keinen Fall bestätigen. Und zwar weder durch ein brüskes Zurückweisen der Frage noch durch eine dem Vorurteil zustimmende Antwort.

### Negativbeispiel

Ein älterer Bewerber sollte dementsprechend auf die Frage *Aren't you too old to cope with all the travel that this job involves?* nicht so antworten: *I didn't think there would be that much. In my old firm I was advised to ›take on the interregional assignments, it will help you to get promotion‹. But nothing came of it. I really only want to travel now in exceptional cases.*

### Kommentar zum Negativbeispiel

Es ist durchaus verständlich, dass der Bewerber nicht mehr in der Gegend herumreisen will – mit seiner Ehrlichkeit schießt er aber ein Eigentor. Selbst wenn er guten Willens ist, die Anstrengungen der Reisetätigkeit auf sich zu nehmen, wird der Personalverantwortliche dieser Antwort entnehmen, dass der Bewerber sich überfordert fühlt. Damit bestätigt der Kandidat Vorurteile gegenüber älteren Bewerbern: Der Personalverantwortliche muss annehmen, dass sich der Bewerber inzwischen zu alt fühlt, um noch überregional eingesetzt werden zu können.

### Antwort-Strategie

Sie sollten den gängigen Klischees vielmehr auf positive Weise entgegentreten: Entkräften Sie Vorurteile, mit denen Sie konfrontiert werden, indem Sie Ihr individuelles berufliches Profil in den Mittelpunkt Ihrer Antworten stellen. Verdeutlichen Sie, dass Sie mit den neuen Aufgaben klarkommen werden, weil Sie über die dafür notwendigen beruflichen Erfahrungen verfügen. Machen Sie ebenfalls anhand von Beispielen klar, dass Sie bereits in der Vergangenheit mit allen Kollegen gut zurechtgekommen sind – mit gewerblichen Mitarbeitern genauso wie mit Akademikern, mit älteren genauso wie mit jüngeren und mit westdeutschen genauso wie mit ostdeutschen Kollegen.

## Positivbeispiel

Besser wäre es, die Frage *Aren't you too old to cope with all the travel that this job involves?* folgendermaßen zu beantworten: *I know that there's travel involved in the job, and I'll be happy to take it on. After all, I've taken on several interregional assignments for my present employer.*

## Kommentar zum Positivbeispiel

Im Vorstellungsgespräch ist es wichtig, nicht an bestimmten Vorurteilen zu rühren. Dies beherzigt der Bewerber diesmal auch in seiner Antwort. Er geht gar nicht auf die Altersfrage ein, sondern führt stattdessen aus, dass er mit den Anforderungen der neuen Stelle vertraut ist. Damit stellt er sein berufliches Profil, zu dem auch bisher schon die Reisetätigkeit gehörte, und nicht sein Alter in den Mittelpunkt seiner Antwort. Diese Taktik wird ihn weiterbringen und die Personalverantwortlichen überzeugen, denn wenn sie den Bewerber wirklich für zu alt hielten, hätten sie ihn gar nicht eingeladen. Letztlich zählt, wie der Bewerber sich selbst einschätzt.

145. Why have you changed employer so often? Do you have difficulty fitting in?

Your answer: _____

_____

_____

_____

_____

_____

_____

_____

_____

_____


146. You've been unemployed for eight months. Will you be able to
cope with the demands of the new job?

Your answer: _____

_____

_____

_____

_____

_____

_____

_____

_____

_____

**Poor answer to question 145**   It wasn't down to me. These days, companies go out of business so quickly. And sometimes I had trouble with my managers.

**Good answer to question 145**   I agree that it isn't ideal. But for each move there were reasons beyond my control. One time the company suddenly went out of business, another time there was a takeover followed by staffing cutbacks. But in every job I was able to accumulate more experience, and in the end that was to my advantage.

**Poor answer to question 146**   I'm sure that with your help I'll be able to get back into a working routine. It isn't that I don't want to. Nobody has given me the chance so far.

**Good answer to question 146**   I'm sure I will. After all, I have several years of professional experience, and I'm well qualified. While I've been looking for work I've kept active and increased my computer knowledge.

147. You finished your degree ten months ago. Why haven't you found a job yet?

Your answer: _____

_____

_____

_____

_____

_____

_____

_____

_____

_____

_____

148. Don't you think that you're too young for this demanding position?

Your answer: _____

_____

_____

_____

_____

_____

_____

_____

_____

_____

**Poor answer to question 147**   I could have got unpaid work experience placements with various companies, but I didn't do all that studying so that I could work for free.

**Good answer to question 147**   I, too, thought it would be easier to get started, particularly as I've already got practical experience. The job market for marketing specialists has become very tight. Because I wanted to find an international employer to make use of my international experience, I only applied to companies with the right profile. That's why it's taken a little longer.

**Poor answer to question 148**   It's interesting that you ask me that. After all, we're always hearing that German graduates are too old. I've made the effort to complete my studies quickly, and now you're telling me that I shouldn't have bothered?

**Good answer to question 148**   No, I don't. I concentrated on relevant study areas and did suitable work placements to prepare for a demanding role. While I was completing my thesis I was also working in a company. I collaborated effectively with older colleagues, and the excellent results we achieved demonstrate that I can handle a project management role.

149. How do you deal with older colleagues who have more experience than you?

Your answer: _____

_____

_____

_____

_____

_____

_____

_____

_____

_____

_____


150. At your age, do you still feel up to the challenges of the job?

Your answer: _____

_____

_____

_____

_____

_____

_____

_____

_____

_____

_____

**Poor answer to question 149**   It is difficult sometimes with older colleagues. I've often found them to be stuck in their ways. I try to work around them, dealing with the new things myself and giving them the impression that nothing has changed.

**Good answer to question 149**   I've always worked well with older colleagues. If you approach them you can benefit a lot from their experience. I always try to instil the same team spirit in older colleagues as in younger ones.

**Poor answer to question 150**   Of course. We older workers have a lot more stamina than the young. We still know what real work is.

**Good answer to question 150**   Because of my professional experience I'm well aware of the challenges I'll face in my new job. I'll give it my all, just as I always have. In recent years I've been involved in restructuring measures which have demanded a lot of personal commitment.

151. You have been abroad a lot. Will you still be able to work here in Germany?

Your answer: _____

_____

_____

_____

_____

_____

_____

_____

_____

_____


152. You have been working in the office, now you want to work in the field. How do we know that you'll cope?

Your answer: _____

_____

_____

_____

_____

_____

_____

_____

_____

_____

**Poor answer to question 151**   Why not? My work abroad has proved that I'm flexible enough, hasn't it?

**Good answer to question 151**   During my assignments abroad I've always maintained my contacts with colleagues in Germany. There were ongoing harmonisation processes which needed to be discussed with the head office in Germany. And, of course, I worked in Germany for many years before I went abroad.

**Poor answer to question 152**   Having seen what my colleagues in the field actually do, I don't think I'll have any trouble with the work.

**Good answer to question 152**   It was a deliberate decision on my part. I've done a lot of cold-calling in my office role. I know how to build customer contacts and look after existing clientele. Of course, I know that I'll be expected to visit a set number of customers each day, and that these visits will sometimes be in the evening. But I'm confident I can handle these duties.

153. Your predecessor in this job came from Leipzig to us in Cologne, just like you. After three months he resigned. Why do you think you won't do the same?

Your answer: _____

_____

_____

_____

_____

_____

_____

_____

_____

_____

_____


154. As a native of Rostock, how do you cope when your colleagues tell jokes about East Germans in the canteen?

Your answer: _____

_____

_____

_____

_____

_____

_____

_____

_____

_____

Are you able to dispel prejudice?   **215**

**Poor answer to question 153**   Cologne or Leipzig – the main thing is that you do your job. I'll manage fine.

**Good answer to question 153**   There are some regional differences in mentality, it's true. But I've relocated several times before without any problems. I think it's important to go to my new colleagues, rather than waiting for them to come to me, and to treat them with respect.

**Poor answer to question 154**   I don't need to put up with that kind of thing. That's what the staff council is for.

**Good answer to question 154**   I'll have a good laugh and probably be able to tell more East German jokes than them. Having a laugh together strengthens team spirit.

155. It says in your CV that you have two children of school age. How will you manage at work if your children are ill?

Your answer: _____

_____

_____

_____

_____

_____

_____

_____

_____

_____

156. You've been out of the workforce for seven years because of your children. Do you think you'll be able to fit back in?

Your answer: _____

_____

_____

_____

_____

_____

_____

_____

_____

_____

**Poor answer to question 155**   I'd rather not think about that. So far they haven't had any real health problems. I hope things will stay that way.

**Good answer to question 155**   I've already decided what I would do in that situation. My mother-in-law can look after the children for a few days, and I also have a good friend who would be able to help out.

**Poor answer to question 156**   I wasn't sure, but my friends tell me that not much has changed in the secretary's office, apart from the fact that it's now called the ›back office‹.

**Good answer to question 156**   I've always stayed in touch with the workplace. For example, by providing holiday cover, but also by talking with former colleagues. Also, I've gone on various courses to update and extend my computer skills and also familiarised myself with internet and intranet applications.

# 14

# What kind of people manager are you?

Dieses Kapitel ist für Sie nur dann interessant, wenn Sie sich um eine Position im Management bewerben. Hier geht es darum, die eigenen Führungsqualitäten zu beweisen. Denn wenn Stellen mit Führungsverantwortung neu besetzt werden, wird die Firmenseite im Vorstellungsgespräch natürlich erfahren wollen, wie es um die Führungsfähigkeiten des Bewerbers bestellt ist.

## Hintergrund

Führungskräfte sollen vorrangig sicherstellen, dass ihre Mitarbeiter Zielvorgaben einhalten. Im Führungsalltag kommt es dabei immer wieder zu Widerständen, Zielkonflikten und anderen Störungen. Im Vorstellungsgespräch will man nun herausbekommen, wie die Bewerber um Führungspositionen derartige Probleme in der Vergangenheit bewältigt haben. Darüber hinaus wird auch überprüft, ob die Bewerber ausreichend Erfahrung mit den unangenehmen Seiten des Führungsalltags haben – wozu beispielsweise Kritikgespräche oder Kündigungen zählen.

## Typische Fehler

Auch wenn man berücksichtigen sollte, dass die Führungskultur je nach Firma variiert, hat sich doch bei der Mehrzahl der Firmen ein persönlich-wertschätzender und zielorientierter Führungsstil durchgesetzt. Deshalb können Führungskräfte, die im Vorstellungsgespräch den Eindruck erwecken, dass sie sich bei aufkommenden Problemen hinter ihrer formalen Position verstecken und bevorzugt durch Anordnungen von oben herab handeln, in der Regel nicht überzeugen. Auch wenn die Antworten vermuten lassen, dass der Bewerber seine künftigen Mitarbeiter nicht motivieren kann, auf gründliche Problemanalysen verzichtet und kein eigenes Engagement bei der Auflösung schwieriger Situationen zeigt, wird er den angestrebten Führungsjob nicht bekommen.

## Negativbeispiel

Führungskompetenz lässt sich mit der Frage *How do you implement management decisions?* überprüfen. Mit der folgenden Antwort fällt der Kandidat leider durch: *I give clear and direct instructions. I avoid unnecessary discussion. At the end of the day, everyone has to recognise that their job depends on the good of the company.*

## Kommentar zum Negativbeispiel

Mit Befehlen und Anordnungen lässt sich heutzutage nicht mehr wirkungsvoll führen. Personalverantwortliche reagieren in der Regel sehr allergisch, wenn sie aus den Antworten von Bewerbern heraushören, dass über die Köpfe der Mitarbeiter hinweg geführt wird. In Zeiten flacher Hierarchien und abteilungsübergreifender Projektarbeit sind selbstherrliche Abteilungskönige ein Störfaktor im Unternehmen. Da Führungsfehler zu Missstimmungen und in letzter Konsequenz auch zu einer hohen Mitarbeiterfluktuation führen, bedeuten sie letztlich Mehrarbeit für die Personalabteilung. Darauf können Personalverantwortliche gut verzichten – und damit auch auf den Bewerber.

## Antwort-Strategie

Um bei den Fragen zur Führungserfahrung zu überzeugen, sollten Sie sich ausreichend Beispiele für gelungene Führungsaufgaben überlegen und auch überzeugende Belege dafür geben können, wie Sie typische Konflikte aufgelöst haben. Lassen Sie erkennen, dass Sie zwar die Zügel in der Hand halten, Ihren Mitarbeitern aber grundsätzlich Wertschätzung und Vertrauen entgegenbringen. Zeigen Sie auf, dass Sie über ein ausgeprägtes Arsenal an Führungsmethoden verfügen. Dazu gehört das erfolgreiche Führen von Mitarbeitergesprächen genauso wie das Einschwören des Teams auf neue Unternehmensziele in Abteilungsmeetings. Geht es in der neuen Stelle auch um Projektleitungen, sollten Sie Beispiele dafür geben, dass Sie auch abteilungsübergreifende Teams zielorientiert führen können.

## Positivbeispiel

Dass ein Bewerber den Führungsalltag überzeugend gestaltet, kann er mit der folgenden Antwort auf die Frage *How do you implement management decisions?* eindrucksvoll belegen: *It's important for employees to understand the reasoning behind a decision. The individual needs to know and understand why I want him or her to do a particular thing. I've always found management by objectives to be an effective approach. I agree with clearly defined objectives for my people – taking their individual potential into account, so that I don't ask too much or too little of them. I monitor their progress with interim reporting. When the opportunity arises, I use meetings to put their achievements into an overall context, so that they can see that their work is actually moving the company forward.*

## Kommentar zum Positivbeispiel

Hier präsentiert sich der Bewerber als erfahrene Führungskraft. Er schafft den schwierigen Spagat zwischen zu hartem und zu weichem Vorgehen. Es wird deutlich, dass er zu jedem Zeitpunkt das Heft in der Hand behält, aber dennoch darauf achtet, dass seine Mitarbeiter ihr individuelles Potenzial einbringen können. Doch dieser Bewerber ist nicht nur ein guter Kommunikator und Organisator, er erhöht auch die Motivation seiner Mitarbeiter, indem er Teilerfolge in Meetings präsentiert. Die Mitarbeiter fühlen sich dadurch ernst genommen und werden bereit sein, auch in Zukunft ihr Bestes zu geben.

157. What management experience do you have?

Your answer: _____

_____

_____

_____

_____

_____

_____

_____

_____

_____

158. What makes for good people management, in your opinion?
     Name three factors.

Your answer: _____

_____

_____

_____

_____

_____

_____

_____

_____

_____

_____

**Poor answer to question 157**   I have natural authority, so I always knew I was destined for management.

**Good answer to question 157**   I've been directly responsible for up to six staff. On projects I've had indirect responsibility for up to eleven additional staff members. So I know from my own experience what's expected of a manager on a daily basis.


**Poor answer to question 158**   Firstly, the ability to assert oneself, secondly, respect and, thirdly, being a good example.

**Good answer to question 158**   In general, I would say: the ability to concentrate energies on a particular goal. I've always found management by objectives to be an effective approach. To do that you have to recognise the potential of your staff. Secondly, you have to distribute tasks appropriately, so as to make the most of this potential. Thirdly, you have to use suitable feedback mechanisms to concentrate their efforts on the objective.

159. What management principles do you apply?

Your answer: _____

_____

_____

_____

_____

_____

_____

_____

_____

160. What positive comments would your present staff make about you? What negative comments would they make?

Your answer: _____

_____

_____

_____

_____

_____

_____

_____

_____

**Poor answer to question 159** I think that humanity, expressed through intuition and empathy, is the key factor in situational management. Strong leadership needs to take a back seat to flexibility. Knowledge of human nature isn't entirely something you can learn, though. You still need a certain amount of natural leadership talent.

**Good answer to question 159** I've achieved good results with management by objectives. Employees appreciate having clear goals to work towards but freedom in how they achieve them. It's also important to back up your staff and get involved yourself, so as to keep things going in the right direction.

**Poor answer to question 160** It would depend on which staff members you asked. There's always a troublemaker in the team. I think most of them would be very pleased with me, a few of them less so, but you have to put up with that as the manager.

**Good answer to question 160** My staff would say that I'm always ready with advice and practical assistance, that I give them sufficient autonomy, and that they can rely on me. Sometimes, they grumble when I want results quickly. But they know that I won't set unattainable goals.

161. Please describe the last conflict you had with a staff member. How did you resolve it?

Your answer: _____

_____

_____

_____

_____

_____

_____

_____

_____

_____


162. How do you motivate your staff?

Your answer: _____

_____

_____

_____

_____

_____

_____

_____

_____

_____

**Poor answer to question 161**   There was one employee who always contradicted me. I tried informal and formal warnings, but in the end I had to sack him. But then he took legal action to get reinstated, and the only thing we could do was sideline him with unimportant tasks.

**Good answer to question 161**   Conflicts are a fact of life. The important thing is how you manage them. My last conflict was with a young employee who felt he was out of his depth, but he just didn't say so. Instead he complained that his colleagues were ignoring him. I spoke with him and with his colleagues, and, in the end, I found him an experienced mentor. After a while, he was able to manage the work by himself.

**Poor answer to question 162**   Motivation is a complex, reciprocal process with many variables. At least, that's what the management books say. It's difficult to find your bearings. At the end of the day, though, what the employee wants is her salary in her bank account at the end of the month. That's motivation enough.

**Good answer to question 162**   I think it's important that my staff always knows why they're doing something. The biggest demotivation is when staff thinks they're being asked to do a pointless task. That's why I hold regular meetings, so I can update my staff on the company's objectives and explain how they can help to achieve them.

163. What reputation does your department/team have in the organisation as a whole?

Your answer: _____

_____

_____

_____

_____

_____

_____

_____

_____

_____

164. How has your management style changed over the years?

Your answer: _____

_____

_____

_____

_____

_____

_____

_____

_____

_____

**Poor answer to question 163**   No particular reputation. We do what has to be done and don't tread on anyone else's toes.

**Good answer to question 163**   A good reputation. The other departments know that we work hard and treat each other fairly. I actually think some of them envy us our good working environment.

**Poor answer to question 164**   I think I've become a bit tired over the years. When I was younger, I was more likely to intervene when I wasn't happy with something. These days, I tend to take a back seat.

**Good answer to question 164**   I've become better at finding a connection with different types of employee. What I might have interpreted as insubordination when I was a young manager, I can now understand much better. Often, it's just insecurity, sometimes, it's overwork, sometimes problems at home. It's rarely a genuine refusal to cooperate. That's why I've got used to a flexible management style, since everyone needs to be handled differently.

165. How do you achieve perfect results from your staff?

Your answer: _____

_____

_____

_____

_____

_____

_____

_____

_____

_____

166. What management qualities does your deputy need in order to cover for you?

Your answer: _____

_____

_____

_____

_____

_____

_____

_____

_____

_____

_____

**Poor answer to question 165** It would be nice if it happened by itself. In reality, though, I have to stand over my staff more than I would like to.

**Good answer to question 165** First of all, I make sure that my staff and I have the same objectives. I ask them to report to me at intervals to make sure they are staying on track. If they aren't, I have time to take corrective measures. That's how I achieve the results I want.


**Poor answer to question 166** They only need to have a rough idea how things work in my department. They don't need to replace me completely. Reliability and honesty are important, of course.

**Good answer to question 166** They have to focus on results, have a good overview of the organisation, have a talent for organising and take the initiative in approaching people. A feeling for the strengths and needs of my staff is also necessary.

167. Have you ever dismissed an employee? How did you go about it, or how would you go about it?

Your answer: _____

_____

_____

_____

_____

_____

_____

_____

_____

_____

168. How do you react when an employee doesn't meet objectives?

Your answer: _____

_____

_____

_____

_____

_____

_____

_____

_____

_____

**Poor answer to question 167**   Sometimes, it can't be avoided. It's an unpleasant task, and I try to get it over with as quickly as possible.

**Good answer to question 167**   Yes, I have had to dismiss staff. Once, staff cutbacks were urgently needed to restore the company's competitive edge. I had one-to-one talks with staff to explain the company's situation. I advised them to promote their strengths aggressively when applying for jobs with other companies. It wasn't an easy task, but I managed to make them understand why the measures were necessary.

**Poor answer to question 168**   Then they're in the wrong job. You have to find out who is responsible for that poor employment decision. The personnel department has to take its share of the blame.

**Good answer to question 168**   With work objectives I always monitor the intermediate goals as well, so I'm usually quick to spot when there's a problem. Then I talk with the staff member to find out the reasons. Is information lacking? Is the task too complex? Is there a lack of know-how? Or is the staff member just trying to get out of an unpopular task? When I've clarified matters, I explain how they can achieve the required objectives in the future.

169. A good staff member gives you the choice: either you increase her salary or she leaves. But your staffing budget is already fully committed. What do you do?

Your answer: _____

_____

_____

_____

_____

_____

_____

_____

_____

_____


170. What experience have you had of project management so far?

Your answer: _____

_____

_____

_____

_____

_____

_____

_____

_____

_____

**Poor answer to question 169**   I promise her that I'll consider a salary increase at the appropriate time. Other than that, I ask her to be reasonable. Times are hard, and we all have to make do without extras.

**Good answer to question 169**   Good staff members don't generally go in for blackmail. However, one solution would be to agree a development plan with the staff member and the HR department. The salary increase could be tied to additional duties. But I wouldn't lay the emphasis on the financial aspect. I would prefer to point out the advantages to the staff member of fast-tracking her career.

**Poor answer to question 170**   It depends on what you mean by project management. I've certainly had experience of some complex tasks. It was hard work, but I would be prepared to take on a project again.

**Good answer to question 170**   I have had responsibility for projects. That includes time management, resource management and, of course, budget control. It was to do with quality measures which I discussed with other departments and implemented. We were able to achieve significant improvements.

171. What in your view are the differences between managing a department and managing a project?

Your answer: _____

_____

_____

_____

_____

_____

_____

_____

_____

_____


172. In what situations do you find it difficult to make a decision?

Your answer: _____

_____

_____

_____

_____

_____

_____

_____

_____

_____

**Poor answer to question 171**   In my department, the staff have to listen to me. In a project group, they can, but they don't have to, which makes the whole thing very difficult. Interdepartmental jealousies are another disruptive influence which makes project management more difficult than managing a department.

**Good answer to question 171**   In a typical project, different departments and areas of the company work together. You have to speak their language, and you have to adapt to the differing needs of the participants. In international project teams, you also have to take cultural differences into account. That makes project management a complex but very interesting task. However, in my department I also adapt to my individual staff members and their personal needs and strengths, so it isn't such a big step from there to project management.

**Poor answer to question 172**   As a manager, I can't afford to be in that situation.

**Good answer to question 172**   Of course it isn't pleasant to have to make decisions based on an uncertain factual basis. But you still have to be able to decide how to move forward. In the final analysis, as a manager, I'm expected to be decisive, and I don't have a problem with that.

# 15

# What are your salary expectations?

Auch wenn im Vorstellungsgespräch zunächst Fragen zu Ihrem beruflichen Profil, zu Ihrer Selbstmotivation oder Ihrer Leistungsbereitschaft im Vordergrund stehen, kommt irgendwann der Punkt, an dem es um Ihre Gehaltsvorstellungen geht. Viele Bewerber scheuen sich vor dieser Frage, da sie befürchten, zu viel zu fordern oder sich unter Wert zu verkaufen. Deswegen zeigen wir Ihnen in diesem Kapitel, wie Sie Ihre Gehaltsvorstellungen realistisch und überzeugend zum Ausdruck bringen können.

## Hintergrund

Manche Bewerber wechseln die Stelle vorrangig deshalb, um einen deutlichen Gehaltssprung zu erzielen. Andere Bewerber wären schon froh, wenn sie in der neuen Stelle noch genauso viel Gehalt bekommen würden wie in dem letzten gut bezahlten Job. Die Firmen haben beim Thema Gehalt ein ureigenstes Interesse daran, gute Kandidaten möglichst günstig »einzukaufen«. Und Bewerber sind natürlich daran interessiert, den eigentlich immer vorhandenen Verhandlungsspielraum möglichst optimal auszunutzen.

## Typische Fehler

Bewerber, die Gehaltsverhandlungen losgelöst von ihrem beruflichen Profil führen, machen es sich unnötig schwer. Schließlich ist eine Gehaltsverhandlung nicht einfach ein Abgleich unterschiedlicher Zahlenkolonnen, sondern die gemeinsame Einschätzung darüber, was der Bewerber für die Firma in der nächsten Zeit leisten wird. Es reicht dabei nicht aus, sich auf den Lorbeeren der Vergangenheit auszuruhen. Frühere Erfolge spielen zwar eine Rolle im Gehaltsgespräch, aber vom Bewerber muss immer wieder herausgearbeitet werden, auf welche Weise er künftig erfolgreich arbeiten wird. Ein weiterer Kardinalfehler ist die Unkenntnis über üblicherweise gezahlte Gehälter in vergleichbaren Positionen. Gewinnt die Firmenseite den Eindruck, dass der Bewerber seinen »Marktwert« nicht kennt, wird ihm unterstellt, dass er auch im späteren Berufsalltag Schwierigkeiten damit haben wird, anspruchsvolle Aufgabenstellungen gründlich vorzubereiten.

## Negativbeispiel

Auf die Frage *Are you worth your salary?* sollten Bewerber nicht überrascht reagieren. Folgende Antwort wäre dementsprechend sehr ungünstig: *I really can't answer that question. It isn't my decision, it's my employer's.*

## Kommentar zum Negativbeispiel

Mit seiner Antwort lässt der Bewerber eine wichtige Chance verstreichen, um zum Abschluss des Gespräches noch einmal sein berufliches Profil ins Spiel zu bringen. Er zeigt sich auch schlecht vorbereitet, denn anscheinend ist ihm nicht bekannt, wie der Gehaltsrahmen für die von ihm angestrebte Position aussieht.

## Antwort-Strategie

Informieren Sie sich vor dem Gespräch über die in Ihrer Branche und für Ihre Position üblichen Gehälter. Hierbei können Sie auf das Internet zurückgreifen. Geben Sie in eine Suchmaschine Ihre Position und das Stichwort »Gehalt« ein. Üblicherweise werden Sie auf eine ausreichende Zahl von Treffern stoßen. Setzen Sie dann in Gehaltsverhandlungen voll auf Ihr individuelles Profil. Liefern Sie Belege dafür, wie Sie Arbeitsprozesse effizienter organisiert haben, Qualitätsverbesserungen herbeigeführt haben, Vertriebsziele erreicht haben, Projekte zum Abschluss gebracht haben, Kollegen eingearbeitet oder vertreten haben oder Sonderaufgaben übernommen haben. Planen Sie bei Ihren Gehaltsvorstellungen auch einen ausreichenden Verhandlungsspielraum ein, damit Sie Ihrem Gesprächspartner etwas entgegenkommen können.

## Positivbeispiel

Vorbereitete Bewerber lassen sich mit der Frage *Are you worth your salary?* nicht aus dem Konzept bringen. So könnte eine überzeugende Antwort aussehen: *I think so. After all, my salary request is in line with the usual remuneration for the position. I have a good grasp of routine marketing support duties, I've already updated catalogues and advertising materials, prepared sales statistics and been responsible for direct marketing campaigns. In addition, I also have experience in organising trade fair participation and promotional events.*

173. What sort of salary do you have in mind?

Your answer: _____

_____

_____

_____

_____

_____

_____

_____

_____

_____

174. How is your present salary structure made up?

Your answer: _____

_____

_____

_____

_____

_____

_____

_____

_____

_____

**Poor answer to question 173**   I'm not really sure. What are you offering?

**Good answer to question 173**   On the basis of my professional experience, I think that a yearly gross salary of 65,000 euros would be appropriate. But I'm prepared to negotiate, because I'm very keen to get this position.


**Poor answer to question 174**   That's an in-house arrangement which I'm reluctant to disclose. Let's say, I get a little more than just my fixed salary.

**Good answer to question 174**   My salary is made up of a fixed and a variable component. On top of that, my company provides me with a mid-range executive car. I'm happy to continue to have my success rewarded by incentives and bonuses.

175. The candidate who sat there before you asked for 20 per cent less than you. Why should we choose you?

Your answer: _____

_____

_____

_____

_____

_____

_____

_____

_____

_____


176. What did you earn in your previous position?

Your answer: _____

_____

_____

_____

_____

_____

_____

_____

_____

_____

**Poor answer to question 175**   I'm the stronger candidate and have excellent qualifications.

**Good answer to question 175**   I think that my present salary level is absolutely justified. After all, on top of my field sales role I've taken on special assignments such as developing sales promotions. The turnover I've achieved has always been well above the average. But I'm prepared to accept an incentive payment arrangement.

**Poor answer to question 176**   I was well paid in my previous job, but times were different then. I suppose I need to be prepared to accept a significant salary cut.

**Good answer to question 176**   I earned 42,000 euros a year. On top of that, my company paid me bonuses, financed professional development courses and provided a direct insurance plan.

177. What would you like to earn with us?

Your answer: _____

_____

_____

_____

_____

_____

_____

_____

_____

_____


178. We definitely can't pay you as much as your present
     employer. So why do you still want the job? Have you been
     asked to resign?

Your answer: _____

_____

_____

_____

_____

_____

_____

_____

_____

_____

**Poor answer to question 177**   Well, if I had the choice, I would ask for more than you're likely to offer. But I'm grateful for any reasonable offer.

**Good answer to question 177**   Up to now, I've earned 62,000 euros gross per year. Now, I'm aiming for a salary of 70,000 euros. I think this is justified, in view of my international experience.

**Poor answer to question 178**   To be honest, things haven't been going well in my firm for a long time. A lot of people have gone already, and I don't want to be the last one left, who turns out the lights.

**Good answer to question 178**   It's important for me to be able to give a job all my energy. The advertised position is a very good match with my professional experience. It's more important to me to work hard with you than to accept a higher salary elsewhere and then find myself marking time. Give me the chance to convince you through my performance.

179. You aren't very good at judging your market value, are you?

Your answer: _____

_____

_____

_____

_____

_____

_____

_____

_____

_____


180. If it was up to me, I would be prepared to pay what you
ask for. But as you know, times are hard. I can't offer you any
more, so what should I do?

Your answer: _____

_____

_____

_____

_____

_____

_____

_____

_____

_____

**Poor answer to question 179**   I really can't comment on that. It's difficult to find out what other people get. But there has to be some compensation for the extra work involved.

**Good answer to question 179**   I'm sorry I haven't been able to convince you yet. The salary I've asked for may be slightly above the usual rate for this position, but you would be benefiting from my self-financed training as a quality auditor. Also, I can conduct negotiations confidently in both English and Spanish, which will be of real benefit to you.

**Poor answer to question 180**   I'm sure that if you look at all the possibilities you can come up with the extra 5,000.

**Good answer to question 180**   Could we agree on a contractual salary increase after six months? Then I could accept the job at the salary you propose and show you that I'm worth the extra money I want.

181. There are several very good candidates who are prepared
     to accept significantly less than you want. Aren't your salary
     demands a bit too high?

Your answer: _____

_____

_____

_____

_____

_____

_____

_____

_____

182. Could you accept 20 per cent less than you want, and then we
     can see how we go after your probation period is over?

Your answer: _____

_____

_____

_____

_____

_____

_____

_____

_____

_____

**Poor answer to question 181**   Yes, I know there's a lot of competition. Do you think I should ask for less? Perhaps my expectations were a bit too high.

**Good answer to question 181**   I think my salary requirements are well justified. They're in the middle of the usual range for this kind of position. And with my experience in customer service I can hit the ground running. I know the industry well, and I have extensive experience of troubleshooting on site.

**Poor answer to question 182**   Yes, if there's no other way.

**Good answer to question 182**   A 20 per cent cut seems like a lot. We should meet each other half way. If on top of that we mention in the contract that my salary will be increased after the probationary period, I'm prepared to accept your offer.

183. OK, we'll offer you a company car for your private use, but you reduce your salary requirement by 10 per cent. How does that sound?

Your answer: _____

_____

_____

_____

_____

_____

_____

_____

_____

_____

184. Aren't your salary demands excessive?

Your answer: _____

_____

_____

_____

_____

_____

_____

_____

_____

_____

**Poor answer to question 183**   I've got my own vehicle, so I don't need a company car. The extra 10 per cent won't bankrupt the company.

**Good answer to question 183**   You see, I'll have to pay tax on 1 per cent of the list price of the company car each month. In my tax situation, that isn't a good deal. In principle, though, I don't have anything against your suggestion. Perhaps we could agree on a 5 per cent reduction and the company car.


**Poor answer to question 184**   I know what I'm worth, and I'm sorry you still haven't recognised that I would be an asset to the company.

**Good answer to question 184**   I see my salary requirement as a reasonable remuneration for my productivity. I can fit into the daily work routine of the department immediately, because I've already worked with the software you use. In financial control I've achieved significant savings for my present employer, and you'll benefit from that as well.

# 16

# What can you expect in the second interview?

Mit der Einladung zum zweiten Vorstellungsgespräch haben Sie eine weitere Hürde im Auswahlverfahren erfolgreich genommen. Jetzt kommt es darauf an, Ihre überzeugende Vorstellung aus dem ersten Gespräch zu untermauern und den Entscheidungsträgern auf der Firmenseite nochmals zu verdeutlichen, dass Sie genau der oder die Richtige für die zu vergebende Stelle sind.

## Hintergrund

Sämtliche Bewerber, die zu einem zweiten Gespräch eingeladen werden, können davon ausgehen, dass die Firmenseite in Runde eins prinzipiell von ihrer fachlichen Kompetenz und ihrem persönlichen Auftritt überzeugt worden ist. In der zweiten Runde wird noch einmal an Punkten nachgehakt, die für die Firma besonders wichtig sind. Hinzu kommt, dass oft zusätzliche Gesprächspartner – beispielsweise künftige Fachvorgesetzte oder Geschäftsführer – neu dabei sind. Der Bewerber darf deshalb in seinen Anstrengungen nicht erlahmen, sondern muss mit seiner Überzeugungsarbeit noch einmal voll durchstarten.

## Typische Fehler

Es kommt häufiger vor, dass gute Bewerber im zweiten Vorstellungsgespräch scheitern, weil sie nicht den unbedingten Willen erkennen lassen, ihre Kompetenz und ihre Erfahrungen voll in die neue Stelle einzubringen. Dies liegt in der Regel nicht an der tatsächlich fehlenden Leistungsmotivation, sondern vielmehr an einer ungeschickten Gesprächsstrategie. Auch wenn der Eindruck entsteht, dass Bewerber nur auf der Suche nach »irgendeiner« Stelle sind, fallen sie bei den Firmenvertretern durch. Problematisch ist es außerdem, wenn von den Bewerbern keinerlei Bezug auf die Inhalte aus dem ersten Vorstellungsgespräch genommen wird. Dann wirkt der Wunsch nach einer künftigen Mitarbeit schnell unreflektiert und bloß aufgesetzt.

### Negativbeispiel

Eine typische Frage im zweiten Vorstellungsgespräch wäre: *Do you still think that you are the right employee for us?* Ungeeignet ist dann diese Erwiderung: *I thought we had covered that in the first interview. I can't say much more than I did then, but I'm still sure.*

### Kommentar zum Negativbeispiel

Die Antwort der Bewerberin ist deutlich zu dünn – etwas mehr hätte die Bewerberin schon ausholen müssen. Für Personalverantwortliche wirkt es befremdlich, wenn Bewerber nach dem ersten Vorstellungsgespräch abschalten und nicht mehr in der Lage oder willens sind, ihre Einstellungsargumente zu wiederholen. Dies wirft ein schlechtes Licht auf das Kommunikationsverhalten im neuen Job, denn es wird auch im Berufsalltag immer wieder Situationen geben, in denen die Gesprächspartner erneut informiert und überzeugt werden wollen. Zudem sind im zweiten Gespräch häufig neue Vertreter der Firmenseite zugegen – und die kennen Ihre Argumente aus dem ersten Gespräch schließlich noch nicht.

## Antwort-Strategie

Nehmen Sie vor dem zweiten Vorstellungsgespräch noch einmal Ihren Lebenslauf und die Stellenausschreibung zur Hand. Reflektieren Sie dann das erste Gespräch (siehe Kapitel *Rückblende: Was hat Ihnen gefallen?* auf Seite 276) und überlegen Sie sich, was für die Firma am wichtigsten ist. Diese Firmenwünsche und -vorgaben sollten Sie von sich aus im zweiten Vorstellungsgespräch ansprechen und anhand von Beispielen begründen, wie Sie diese Aufgaben erfüllen werden. Wenn Sie auf neue Gesprächspartner treffen, sollten Sie auf jeden Fall eine kurze Selbstpräsentation Ihres Werdeganges liefern (siehe Kapitel *Why should we give you the job?* auf Seite 22). So geben Sie dem Gespräch Dynamik und liefern geeignete Ansatzpunkte für den weiteren Verlauf. Auch Randfragen wie Kündigungsfristen und Umzugspläne sollten Sie vor dem zweiten Gespräch klären, um zu zeigen, dass Sie den neuen Job auch wirklich wollen.

## Positivbeispiel

Um auch im zweiten Vorstellungsgespräch zu punkten, sollte die Frage *Do you still think that you are the right employee for us?* beispielsweise so beantwortet werden: *I very much enjoyed our last interview. It reinforced my wish to work for you as a travel consultant. I'm experienced both in preparing offers and in processing and monitoring bookings. I also have extensive experience of looking after and soliciting corporate clients. You emphasised in our last conversation that it's important to you to expand your corporate business. I'd like to assist you in that and give you the benefit of my experience.*

## Kommentar zum Positivbeispiel

Im zweiten Vorstellungsgespräch ist es wichtig, herauszuarbeiten, dass die Entscheidung, zu einem neuen Arbeitgeber zu wechseln, bewusst getroffen worden ist. Der Bewerberin gelingt das mit dieser Antwort sehr gut. Sie verweist auf das erste Gespräch und arbeitet die Schnittmenge von bisherigen und zukünftigen Aufgaben heraus. Mit dem Verweis auf eine Information, die sie im ersten Gespräch erhalten hat, verdeutlicht sie die Ernsthaftigkeit ihrer Entscheidung, die intensive Auseinandersetzung mit der Stelle und den Nutzen, den die Firma von einer Einstellung hat. Die Bewerberin wird die neue Firma beim Ausbau des Firmenkundengeschäftes maßgeblich unterstützen können, also wird auch die Firma an ihrer Einstellung ein großes Interesse haben.

185. How did you feel the last interview went?

Your answer: _____

_____

_____

_____

_____

_____

_____

_____

_____

_____

186. Is there something you didn't tell us in the first interview which we need to know?

Your answer: _____

_____

_____

_____

_____

_____

_____

_____

_____

**Poor answer to question 185**  I was quite flustered, I'm not sure I was able to get everything across that I wanted to. Were you happy with how it went?

**Good answer to question 185**  I found our last interview very constructive. It reinforced my wish to work for you.

**Poor answer to question 186**  Well, we didn't really talk about my leisure interests. I love good literature and classical music. My wife and I have an annual subscription for the state opera.

**Good answer to question 186**  When you described my future role to me, it became obvious that there were a lot of overlaps with my present work. I'd like to emphasise again that I've successfully carried out focused direct marketing campaigns. The campaigns I initiated had a high take-up rate. I'm familiar with the particular situation of medium-sized businesses. I've outsourced marketing operations, implemented partnership alliances and managed external service providers.

187. How are your job applications going?

Your answer: _____

_____

_____

_____

_____

_____

_____

_____

_____

188. Please describe your career development again for the benefit
of our new interviewer. Which aspects are important for the
vacant position?

Your answer: _____

_____

_____

_____

_____

_____

_____

_____

_____

**Poor answer to question 187**  I thought it would be easier. Someone must have a use for my hard work. At the moment, though, it's going very slowly.

**Good answer to question 187**  I've targeted my applications carefully. I've concentrated on positions which show a close match with my profile. I was very pleased to be invited to this interview, because I really identified with your job description.

**Poor answer to question 188**  Yes, of course. After school I wasn't sure what I wanted to do. I took my parents' advice and did an apprenticeship in banking. Then I did my national service, which I really enjoyed. After that I studied business administration. I remember my work placements clearly, they were formative experiences. Then I was lucky to get my first job, my professor had good connections. Thanks to my professional experience a headhunter approached me and placed me with my present employer. Now things aren't going so well, so I'd like to start with you.

**Good answer to question 188**  I have several years of managerial experience in HR. Of course, I keep my knowledge of labour law, social security, taxation and tariffs continuously up-to-date. As well as the usual administrative functions in HR I was also responsible for HR marketing and HR procurement. For my last employer I introduced performance standards in employment contracts and pay scales and supervised human resource development. I'd like to use this wide-ranging HR experience in your company.

189. What makes you so sure that you're suited to the position?

Your answer: _____

_____

_____

_____

_____

_____

_____

_____

_____


190. You've had time to digest our last interview. What has strengthened your determination to work for us, and what hasn't?

Your answer: _____

_____

_____

_____

_____

_____

_____

_____

_____

**Poor answer to question 189**   I came away from the last interview with a good feeling. So I really thought that we were in agreement about my suitability for the job.

**Good answer to question 189**   The primary reason was that there are so many overlaps between my past role and the one you described to me. As an export coordinator I've dealt with the complete process from preparing an offer to arranging delivery. In complaint processing I've been responsible for clarification of both technical and commercial issues. Additionally, I've participated in international projects in Eastern Europe and prepared the relevant agency agreements. I've informed myself thoroughly about your company and the markets you serve, and our first interview confirmed my impression that I would be a good fit for your company.

**Poor answer to question 190**   I would definitely like to work for you. Sometimes, though, I felt a bit unsure and couldn't quite see what your questions were getting at. All in all, though, I'd say that I'm a good fit for your company.

**Good answer to question 190**   Actually, everything that was said confirmed my decision. Your detailed explanations about the position were exactly in line with the job advertisement and my profile. I felt the interview itself was very friendly and productive. So I'm completely sure now that I want to work for you.

191. What do you expect from this second interview?

Your answer: _____

_____

_____

_____

_____

_____

_____

_____

_____

_____


192. Looking back to our last interview: name three points that
     speak for you as a candidate.

Your answer: _____

_____

_____

_____

_____

_____

_____

_____

_____

_____

**Poor answer to question 191**   I'm not sure. I really just want to sign the employment contract.

**Good answer to question 191**   I would hope for a confirmation of the outcome of the first productive interview. Also, I'm very pleased to have the opportunity to meet my new line manager. Perhaps I'll also get the chance to have a look around the department?


**Poor answer to question 192**   Firstly, there's my above-average motivation, secondly, my teamwork skills and, thirdly, my communicative personality.

**Good answer to question 192**   Firstly, I've always invested in my professional development outside my work hours. So I was able to use my knowledge from computer courses to help my colleagues learn new software. Secondly, my ability to liaise well with others has always helped me in my production work. Because of that, I was able to help my team achieve optimal set-up times when retooling. Thirdly, I've taken part in special assignments alongside my work on the production line. In a cost reduction workgroup I worked together with colleagues from development, financial control and purchasing to achieve economies.

193. What does your wife/husband/partner think about your wanting to work for us?

Your answer: _____

_____

_____

_____

_____

_____

_____

_____

_____

_____

194. What's your family's attitude to the relocation that will inevitably come with this position?

Your answer: _____

_____

_____

_____

_____

_____

_____

_____

_____

_____

**Poor answer to question 193**   We don't discuss those kinds of issues.

**Good answer to question 193**   I've talked with my partner about the job. Just like me, he thinks that the new position is a good fit with my professional experience. He'll be as delighted as me if I'm successful.

**Poor answer to question 194**   You don't have a choice these days, you have to go where the jobs are.

**Good answer to question 194**   I discussed my plans with my family at an early stage. We've already taken a look around Munich. My family like, it here as much as I do. Also, we've relocated to a new town once before, and therefore we're confident we'll have no trouble settling in this time.

195. If you compare your present job with the new position: in what areas do you lack experience?

Your answer: _____

_____

_____

_____

_____

_____

_____

_____

_____

_____

196. When could you start work for us?

Your answer: _____

_____

_____

_____

_____

_____

_____

_____

_____

_____

**Poor answer to question 195**   Well, to be honest I'll need your help with a few things at first. But after all, it's normal that there are new things to learn in a new job. I'll get used to my new duties in the end.

**Good answer to question 195**   I already have experience in all the key areas of the new role. I can draw up practicable concepts, co-ordinate liaison with the development teams and ensure an optimal use of resources. As I'm already familiar with internet applications, it won't take me long to familiarise myself with web portal design. Although I haven't had direct subordinates in the past, I have been responsible for leading project groups of up to eight colleagues.

**Poor answer to question 196**   I'd have to hand in my resignation first. Then, I think, there's three months' notice from the end of the month, or something like that. But I'd also like to go on holiday for a month, to have some quality time with my family for once.

**Good answer to question 196**   I have to give three months' notice. If you needed me sooner than that, I'd approach my employer and try to find a way of being released from my contract more quickly. I don't think they would be obstructive.

197. What short-term and medium-term career goals will you pursue with us?

Your answer: _____

_____

_____

_____

_____

_____

_____

_____

_____

198. What urgent questions do you have as a result of your last interview?

Your answer: _____

_____

_____

_____

_____

_____

_____

_____

_____

_____

**Poor answer to question 197**   Well, to get the job first of all, then to settle in, then to look for opportunities for promotion or at least a salary increase.

**Good answer to question 197**   In the short term, I want to become good enough at the job to be a real help to my colleagues. In the medium term, I'd be interested in taking on extra duties. If I can prove myself by carrying out special assignments successfully, then I'd be interested in taking on a more responsible role after a while.


**Poor answer to question 198**   I'd like to know more about the holiday arrangements. Can I carry leave over to next year? Then there's overtime – is that paid, or do I take time off in lieu? And you haven't really explained the bonus scheme to me yet.

**Good answer to question 198**   I'd like to know who I'll be working with when it comes to interfacing with other areas of the business. Is there a regular system, or are new project groups formed for each task? Also, I'd like to know more about the make-up of the department. Who will be working with me, and what are my colleagues' specialist backgrounds?

199. When will you tell your present employer that you intend to leave?

Your answer: _____

_____

_____

_____

_____

_____

_____

_____

_____

_____

200. You and two other candidates are in the final selection. Why should we choose you and not one of the others?

Your answer: _____

_____

_____

_____

_____

_____

_____

_____

_____

_____

_____

**Poor answer to question 199**  They've known for a long time that I'm going to leave. So I don't need to tell anybody.

**Good answer to question 199**  I'll tell my employer as soon as you've given me the green light. I still have to finish off some tasks and pass on the results. As they'll have to find a successor for me, I'd like to give my employer plenty of notice.

**Poor answer to question 200**  You should pick me, because I'm the right person for the job and you definitely won't regret your decision.

**Good answer to question 200**  Of course, I can only speak for myself. I want to take on the role, because I have a lot of experience in this area. As sales team assistant I've updated catalogues and advertising material. I've built up and maintained contacts with the trade press. I've presented my company's products at trade fairs. Through special events I've significantly increased product recognition amongst our target groups. I've also been involved in translating media and market research data into marketing and sales campaigns. I want to take on the new position to make use of this experience.

# 17

# What questions should you ask?

Im Vorstellungsgespräch werden Ihnen nicht nur Fragen gestellt – auch Ihre eigenen Fragen sind wichtig. Wir haben Sie bereits darauf hingewiesen, dass ein Vorstellungsgespräch erst dann erfolgreich läuft, wenn es nicht zum Frage-Antwort-Spiel verkommt, sondern sich zu einem echten Dialog entwickelt. Dazu gehört auch, dass Sie Fragen stellen dürfen. Mit den richtigen Fragen können Sie nochmals Ihr Interesse an der Stelle unterstreichen.

### Ihre Fragen bitte

Überlegen Sie sich Ihre Fragen auf jeden Fall vor dem Gespräch, denn sonst kann es bedingt durch den Stress des Vorstellungsgespräches passieren, dass Ihnen plötzlich gar nicht mehr einfällt, was Sie eigentlich fragen wollten. Notieren Sie Ihre Fragen deshalb auf einem Blatt Papier, das Sie im Gespräch dabeihaben. Anregungen für Ihre Fragen finden Sie in der folgenden Übersicht:

- How big is the team I'll be working with?
- How many employees will I be responsible for?
- What will my induction be like?
- Who is my line manager?
- Is there an organigram of the company?
- Could I have a look at my workplace?
- Is this a newly-created post?

- If not: how long was my predecessor in this post?
- How is the post integrated into the company structure?
- Which departments will I be working with most closely?
- Which departments/managers will I be reporting to?
- How much time is allotted to each of my duties?
- How big is the travel component of this position?
- Will I also be working abroad for the company?
- Are there any professional development opportunities?
- Are there opportunities for promotion?
- Are there any particular company benefits?
- Is flexitime available?
- Is there time off in lieu of overtime?
- What is the annual leave entitlement?
- How high is the salary, and how is it structured?
- Are there fringe benefits? A company pension scheme/life insurance plan?

Sie können Ihre Fragen stellen, wenn Sie merken, dass Sie sich in einer nicht so strukturierten Phase des Vorstellungsgesprächs befinden. Üblicherweise wird man Sie im letzten Drittel des Gesprächs auch auffordern, eigene Fragen zu stellen. Achten Sie darauf, zunächst Fragen zu den neuen Aufgaben, zur Einarbeitung, zu den neuen Kollegen oder dem neuen Vorgesetzten zu stellen. Fragen zu den Urlaubstagen, zu Sozialleistungen, zur Gleitzeit oder zum Gehalt gehören an das Ende des Gesprächs (Fragen zum Gehalt finden Sie in dem Kapitel *What are your salary expectations?* auf Seite 239). So zeigen Sie, dass Sie nicht vornehmlich am Gehalt Interesse haben, sondern vor allem an der ausgeschriebenen Stelle.

# 18

# Rückblende: Was hat Ihnen gefallen?

Wenn Sie im Auswahlverfahren so weit gekommen sind, dass Sie zwei (oder mehr) Vorstellungsgespräche mit den Vertretern der Firmenseite geführt haben, sind Sie fast am Ziel. Die Wahrscheinlichkeit, dass man Ihnen einen neuen Arbeitsvertrag anbietet, ist jetzt sehr hoch. Auf Seiten der Firma werden die zwei bis drei Kandidaten, die es bis in die letzte Runde geschafft haben, noch einmal einem Vergleich unterzogen: Was spricht für eine Einstellung? Was dagegen? Wo liegen Chancen? Welche Risiken sind zu bedenken?

## Ziehen Sie eine Zwischenbilanz

Auch Sie sollten zu diesem Zeitpunkt – genauso wie die Firmenseite – eine kritische Zwischenbilanz ziehen. Schließlich geht es um eine wichtige Entscheidung, und die sollte nicht einfach »aus dem Bauch heraus« getroffen werden. Gehen Sie die bisherigen Gespräche in Gedanken noch einmal vom Anfang bis zum Ende durch. Überlegen Sie sich, was Sie überzeugt hat, mit welchen Bedingungen Sie leben können und in welchen Bereichen es Schwierigkeiten geben könnte. Vergleichen Sie die neue Stelle mit Ihrer momentanen und überlegen Sie, was sich verbessern, verschlechtern oder was gleich bleiben würde. Die folgenden Fragen helfen Ihnen dabei, eine gründliche Zwischenbilanz zu ziehen und eine Entscheidung zu treffen:

- Entspricht die neue Stelle meinen Erwartungen?
- Werde ich mit den Anforderungen der neuen Stelle zurechtkommen?
- Wo sehe ich Schwierigkeiten?
- Welche Erfahrungen kann ich einbringen?
- In welchen Bereichen muss ich noch dazulernen?
- Ist mein Arbeitsplatz / Büro ansprechend ausgestattet?
- Komme ich mit dem neuen Chef klar?
- Gibt es einen Ansprechpartner für die Einarbeitung?
- Welchen Eindruck haben die Kollegen auf mich gemacht, die ich bisher kennen gelernt habe?
- Ist die Stimmung in der Firma konstruktiv?
- Gibt es Entwicklungsmöglichkeiten in der neuen Stelle?
- Welchen Ruf hat die Firma in der Branche?
- Wie sicher ist der neue Arbeitsplatz?
- Stimmt die Bezahlung?

Letztlich wird es die perfekte Stelle nur in Ausnahmefällen geben: Mit dem einen oder anderen Kompromiss werden Sie leben müssen. Dennoch sollten Sie sich nicht auf einen faulen Kompromiss einlassen, sodass Sie nach kurzer Zeit wieder vor den alten Problemen stehen. Wägen Sie Vor- und Nachteile also gründlich ab, um dann eine Entscheidung für oder gegen die neue Firma zu treffen.

# 19

# Formulierungen, die Ihnen weiterhelfen

In unseren 200 gelungenen Beispielantworten haben Sie zahlreiche Formulierungen kennen gelernt, die Ihnen auch in Ihrem englischen Job-Interview dabei helfen werden, sich in ein gutes Licht zu rücken. Im Folgenden werden wir Ihnen noch ein paar zusätzliche Formulierungen präsentieren, damit Sie in jeder Situation die passenden Vokabeln parat haben.

## Etwas Smalltalk bricht das Eis

Im Mittelpunkt eines Job-Interviews stehen natürlich Ihr individuelles berufliches Profil und Ihre Persönlichkeit (Soft-Skills). Es kommt darauf an, möglichst viele Überschneidungen zwischen Ihrem Profil und dem Anforderungsprofil des Unternehmens deutlich zu machen. Aber bevor es richtig mit den Fragen losgeht, wird zur Auflockerung gern etwas Smalltalk betrieben. Wie auch in anderen Gesprächen im Businessalltag kommt man nicht gleich zur Sache, sondern möchte das Eis etwas brechen und sich einen ersten persönlichen Eindruck verschaffen.

Wichtig für Sie dabei ist, bereits beim Smalltalk auf eine positive Gesprächsatmosphäre hinzuarbeiten. Dies gelingt Bewerbern natürlich nicht, wenn sie sich einsilbig geben und schon zu Gesprächsbeginn nach (englischen) Worten ringen. Nutzen Sie die Chance des Smalltalks, um in der Fremdsprache erst einmal etwas warmzulaufen und sich von Anfang an souverän zu präsentieren.

# Beispielformulierungen Smalltalk

»How was your journey?«

Your answer: _____

_____

_____

_____

_____

Good answer: »Fine, thanks. It was good to get to know (place). It was easy to get here from the airport, too.«

»Did your flight go smoothly?«

Your answer: _____

_____

_____

_____

_____

Good answer: »Yes, thanks. I flew with (airline). There are several direct flights a day, so it gives you a lot of flexibility time-wise.«

»What's the weather like back home?«

Your answer: _____

_____

_____

_____

_____

Good answer: »Our spring/summer/autumn/winter has its own special charm. At the moment it's a bit cooler/warmer than usual at this time of year.«

»Is this your first time in (place)?«

Your answer: _____

_____

_____

Good answer: »Yes, it is. It's nice to have the chance to see (place).«
/»No, actually it isn't. I've been here a few/several times on business.
I know my way around (place), and it's good to be back here again.«

»Have you had the chance to get a first impression of the company?«

Your answer: _____

_____

_____

_____

Good answer: »Yes. I was made to feel very welcome. They were very
helpful at reception and made sure that I found my way here without
any problems.«

## Professionell bis zum Schluss

Sorgen Sie dafür, dass Sie auch am Gesprächsende einen guten Eindruck hinterlassen. Der Smalltalk zum Schluss hat die gleiche Funktion wie der zu Beginn des Job-Interviews: Sie signalisieren damit der Firmenseite, dass Sie ein umgänglicher Teamplayer sind. Am Gesprächsende bietet es sich an, den weiteren Ablauf des Auswahlverfahrens anzusprechen. So signalisieren Sie noch einmal ehrliches Interesse an der zu vergebenden Stelle. Manchmal bietet es sich auch an, noch ein Zusatzargument zu bringen, das der Firmenseite eine positive Entscheidung für Sie erleichtern könnte. So könnten Sie an dieser Stelle auch Referenzen anbieten, vorausgesetzt, diese wurden vorher noch nicht von Ihnen eingefordert.

### Beispielformulierungen für die Schlussphase

»I'd like to put forward Mr/Ms (name), (position) at (company), as my personal and professional referee.«

»Are there any other documents you need from me?«

»How soon can I expect to hear from you?«

»Will there be any further stages in the selection process?«

»The detailed information you've given me has strengthened my interest in the position. I'd love to be in your team.«

»I'd like to thank you for this constructive conversation, and I'd be delighted to start work with you.«

## Zusätzliche Formulierungen

Auch bei bester Vorbereitung kann es Ihnen passieren, dass Ihnen im Job-Interview vor Stress der Gesprächsfaden einmal reißt. Dann ist es gut, wenn Sie auf Formulierungen zurückgreifen können, die Sie wieder ins Gespräch zurückbringen. Wir haben daher für Sie noch einige Formulierungen zusammengestellt, mit denen Sie das Gespräch auch in kniffligen Situationen schnell wieder auf Ihr Stärkenprofil zurückführen können.

## Weitere Beispielformulierungen für das Job-Interview

»My duties to date have included ... and ...«
(»Zu meinen bisherigen Aufgaben gehörte ... und ...«)

»The main emphasis of my work is ... and ...«
(»Schwerpunkte meiner Arbeit sind ... und ...«)

»In recent years I've systematically developed my knowledge of ... , ... and ...«
(»In den Bereichen ..., ... und ... habe ich meine Kenntnisse in den letzten Jahren gezielt ausgebaut.«)

»In my university studies / training / professional development I specialised in ..., ... and ...«
(»Im Studium / der Ausbildung / der Fortbildung habe ich mich auf die Bereiche ..., ... und ... spezialisiert.«)

»I was involved in the ... and ... projects.«
(»Ich war an den Projekten ... und ... beteiligt.«)

»My responsibilities include ... and ...«
(»Mein Verantwortungsbereich umfasst ... und ...«)

»My particular strengths are in the areas of ... and ...«
(»Besondere Stärken habe ich in den Bereichen ... und ... «)

»In my last job I dealt with ..., ... and ...«
(»Auch bei meinem letzten Arbeitgeber habe ich die Aufgaben ..., ... und ... bearbeitet.«)

»In the short term / medium term, I'd like to take on responsibility for ...«
(»Kurzfristig / Mittelfristig möchte ich Verantwortung als ... übernehmen.«)

»I'm keen to make use of my experience in this new position.«
(»Meine Erfahrungen würde ich gerne bei Ihnen einsetzen.«)

# Mehr Sicherheit für den doppelten Stresstest

Mit Ihrer intensiven Vorbereitung auf englische Job-Interviews haben Sie ein hartes Stück Arbeit hinter sich gebracht. Allerdings haben Sie sich mit Ihrem Einsatz gleich mehrere Vorteile verschafft. Ein wesentlicher Aspekt Ihrer Vorbereitungsleistung ist darin zu sehen, dass Sie nun über umfangreiche Praxiskenntnisse in »Karriere«-Englisch verfügen. Sie kennen jetzt das spezielle Vokabular, das Ihnen dabei helfen wird, Ihre beruflichen Stärken auch auf Englisch anderen gegenüber plausibel zu vermitteln. Aber nicht nur das, Sie sind jetzt auch mit den besonderen Fragenstellungen in Job-Interviews vertraut. Gerade die Beantwortung der in Job-Interviews zentralen Frage *Why should we give you the job?* wird Ihnen nun keinerlei Schwierigkeiten mehr bereiten. Im Gegenteil, Sie können nun den Nutzen Ihrer künftigen Mitarbeit für das Unternehmen in Ihrer Antwort herausstellen und machen sich auf diese Weise zum gefragten Wunschkandidaten. Der doppelte Stresstest, ein Job-Interview auf Englisch, hat für Sie damit seinen Schrecken verloren.

Auch bei Ihrer weiteren beruflichen Entwicklung wird es Ihnen helfen, wenn Sie bei passenden Gelegenheiten ein paar flüssige Sätze auf Englisch zu Ihren beruflichen Aufgaben, zu besonderen Projekterfolgen und zu persönlichen Stärken äußern können. Schließlich wird Ihre Karriere mit dem Bestehen des nächsten Job-Interviews noch lange nicht zu Ende sein. Sie zählen zu den gefragten Leistungsträgern, für die sich immer wieder neue Karriereoptionen ergeben werden. In anspruchsvollen Arbeitsfeldern mit länderübergreifenden Unternehmensstrukturen wird Englisch als Geschäftssprache künftig noch wichtiger werden, als es das heute schon ist. Und weltweite Geschäftskontakte lassen sich bei Bedarf auch als Karrierekontakte nutzen, vorausgesetzt, Sie beherrschen Ihr Selbstmarketing auch auf Englisch.

Wenn Sie sich über weitere Ratgeber und unsere persönlichen Beratungsleistungen informieren möchten, finden Sie unsere Angebote unter *www.karriereakademie.de*. Als kompetente Ansprechpartner stehen wir Ihnen in den Themenbereichen Job-Interview, Bewerbungsunterlagen, Assessment-Center, Karrierestrategie, Arbeitszeugnis und Rhetorik als Berater und Coaches zur Verfügung.

Für Ihre englischen Job-Interviews wünschen wir Ihnen viel Erfolg! Good luck!

*Christian Püttjer & Uwe Schnierda*

# Register